A THEORY OF MARKET STRATEGY

A THEORY OF MARKET STRATEGY

Richard J. Geruson

New York   Oxford
OXFORD UNIVERSITY PRESS
1992

Oxford University Press

Oxford   New York   Toronto
Delhi   Bombay   Calcutta   Madras   Karachi
Kuala Lumpur   Singapore   Hong Kong   Tokyo
Nairobi   Dar es Salaam   Cape Town
Melbourne   Auckland

and associated companies in
Berlin   Ibadan

Published by Oxford University Press, Inc.,
200 Madison Avenue, New York, New York 10016

Oxford is a registered trademark of Oxford University Press

Library of Congress Cataloging-in-Publication Data

Geruson, Richard J.
A theory of market strategy / Richard J. Geruson.
p.   cm.   Includes bibliographical references and index.
ISBN 0-19-506800-9
1. Computers—Marketing.   I. Title.
HD9696.C62G47   1992   004'.068'8—dc20   91-7939

9 8 7 6 5 4 3 2 1

Printed in the United States of America
on acid-free paper

To Nanette

Preface

Business strategy has two sides: the demand side and the supply side. The supply side is primarily about cost-based advantages like scale economies, experience curves, and shared activities. This book is about the other side of business strategy, the demand side. I have labeled this subfield *market strategy*. Just as markets are places where products and customers come together through the mechanisms of price and information, market strategy is about how product behavior, customer characteristics, pricing techniques, and market information can be used to strategic advantage.

Success in market strategy comes from different sources. But I propose that the essence of effective market strategy can be distilled and articulated in one principle: the exploitation of *two-dimensional market heterogeneity*. This is a new idea that summarizes the strategic logic of different types of market behavior. It is a unifying principle for integrating into one framework the fundamental sources of effectiveness in market strategy. These sources include forms of behavior that can be roughly described as enhanced versions of product differentiation, price discrimination, and market segmentation. These sources also include behavior such as information signaling and the reduction of customer-perceived purchase risk that exploit inadequate market information, an implication of two-dimensional market heterogeneity.

The book draws on ideas from a variety of disciplines, including traditional subfields in economics, marketing, and business strategy but also nontraditional areas such as spatial economics, the economics of information acquisition, and the psychology of customer purchase risk. The result is a new framework useful to any business that wants to develop, understand, or diagnose market strategy.

The second half of this book shows how this framework can be applied to understand actual market behavior and strategy in the mainframe computer business. In-depth examples from IBM, Cray Research in the supercomputer niche, Tandem Computer in the fault-tolerant computer niche, Control Data Corporation, and others, show how the ideas of the framework developed in the first half of the book can be implemented in the real world. Such detailed examination of observed behavior also reveals the hidden sources of success (and failure) of IBM and its best competitors. Many lessons are learned from these masters of market strategy when their secrets are unlocked with the key of two-dimensional market heterogeneity.

Orange, CA   R. J. G.
March, 1992

Acknowledgments

A few people made suggestions that improved the content or helped shape the direction of this book. Derek Morris and John McGee of Oxford University suggested I incorporate ideas in the area of spatial competition. Ben Shapiro of Harvard Business School raised my awareness of the importance of risk reduction in understanding IBM's success. Bob Dolan, also of Harvard Business School, led me to literature on price segmentation and its potential for bridging disciplines.

My father, Richard T. Geruson, helped with the structure and expression of ideas, and my mother, Joan T. Geruson, provided encouragement until the end of the project that helped assure its completion. My editor, Herb Addison, suggested generic guidelines for rewriting the first chapter, which was improved substantially as a result. Former employers IBM and McKinsey & Co. (and particularly Robert O'Block, McKinsey's Boston office director) recognized the potential value of this undertaking and provided leaves of absence needed to write the bulk of this book. Most importantly, I am thankful to my wife, Nanette Fondas, whose support, patience, and constructive intellectual critiquing helped at all stages of this project, from its beginning to its end.

The written material I drew upon was vast: from legal proceedings of IBM antitrust trials and statistical data bases of customer purchasing patterns, to economic models of horizontal differentiation and the historical development of marketing theory. Accumulating such far-reaching intellectual debt is an inevitable outcome of constructing an original, cross-disciplinary framework. But the debt to authors I wish to recognize by name is not so much intellectual as motivational.

An obscure article by T. V. Atwater in the late 1970s provoked thinking on my part that eventually led me to the works of Wendell Smith and Wroe Alderson in the late 1950s, a time unparalleled in its integrative perspective on the fields of marketing and economics. And although the ideas of that era have not borne fruit, the holistic approach was an important inspiration for this modern effort.

Contents

Figures

Tables

A THEORY OF MARKET STRATEGY

1

Introduction

In the beginning, God created light. . . . Then, in His own image and likeness, God created man. But man, ever striving to satiate his appetites, bit the forbidden fruit and lost paradise. No longer able to fulfill his desires through divine benevolence, man created the "market," and "market behavior" became the driving mechanism for satisfying human wants in the physical world.

The phenomenon of market behavior, in part, defines the genesis of human civilization. But only since the industrial revolution, did its role become so prominent as to attract intellectual scrutiny. To Adam Smith, it was guided by an invisible hand. To Karl Marx, it was a pattern of oppression. To Thomas Jefferson, it was the logical compliment to democracy. And now, stripped of its ideological badges, market behavior is ranked above political action and military tactics as the real source of global power. Its vigor is a key to the success of Western economies; its absence, a cause for the disintegration of the Soviet Union.

To understand this phenomenon that lies at the foundation of civilization, it is necessary to conduct an excavation. And in many ways this document contains the results of that excavation: a skeleton—not a physical skeleton unearthed with a pick and shovel, but an intellectual skeleton revealed through my research at Oxford University and Harvard Business School and through my observation at IBM and many Fortune 500 companies served by the consulting firm McKinsey & Co.

After sifting through the jumbled surface layers of seemingly unrelated market behaviors and theoretical concepts, an underlying common logic and overriding principle emerges that reveals the basic components of market strategy and how they can be integrated into a singular framework. The result is a guide and tool for businesses, a skeleton upon which the substance of market strategy can be fleshed out. The backbone of this skeleton is the new idea of "two-dimensional market heterogeneity," the unifying principle and most important concept for understanding the anatomy of market strategy.

The essence of strategy lies in the relationship of a firm to its environment. The essence of a market lies in the bringing together of products and customers. Market strategy is thus the relationship of a firm's product (and price) behavior to its customer environment. As such, the study of market strategy is not confined to one disciplinary perspective but lies at the intersection of two disciplinary areas: economics and marketing.

In economic terms, it is focused on the demand side of strategy. In marketing terms, it is concerned with the customer side of strategy. Because of this focus on the demand side and concern with the customer environment, "market strategy" has a more specific meaning than "business strategy" or the even broader concept of "corporate strategy." Similarly, it has a different meaning from "competitive strategy," which focuses on the competitive environment.

This cross-disciplinary approach not only facilitates interdisciplinary bridge building but leads to the development of a specific framework that could not otherwise be developed if the study were confined to either one area or the other. This framework combines the strengths of each area—the logic of economics and the customer perspective of marketing—and integrates previously separate concepts through a common underlying logic.

This logic however, is not the logic of industrial organization (IO) economics upon which the competitive strategy literature of the past decade is built. Rather, it is the more fundamental microeconomic logic of demand curve analysis. The rigorous logic captured in the elegantly simple construct of the demand curve is a powerful heuristic for understanding the customer side of strategy because it is constructed from the building blocks of "utility," the satisfaction customers derive from products.

The sources of advantage on the supply side of strategy have been clearly identified and extensively examined. For example, IO-based strategy frameworks are effective in illuminating cost-based sources of strategic advantage with such concepts as *economies of scale* and *economies of scope*. Consulting frameworks such as the growth/share matrix and the product portfolio are similarly useful in understanding supply-side sources of advantage with the parallel concepts of *experience curves* and, more recently, *shared activities*. But even the promotors of such frameworks now recognize that "the grand flow of strategic advantage in business is at a turning point. . . . Competitive advantage based on cost is going through a transition to advantage based on [customer] choice."[1] And the basis for that choice is greater variety and value, or in short, differentiation. Yet the dynamics of differentiation, the different dimensions of differentiation, the anatomy of demand-side sources of strategic advantage are never clearly delineated; rather they are lumped together in the abused, catchall category of "differentiation."[2]

This study pries open the black box of "differentiation" and peels the onion of "market heterogeneity" to identify the dimensions of each and their implications for market strategy. Thus, in behavioral terms, this book is about differentiation, whether it be quality based, variety based, or price based. In structural terms, this book is about the exploitation of "customer heterogeneity," whether it be based on universal customer benefits, segment-specific customer benefits, or customer price sensitivities.

ANATOMY OF MARKET STRATEGY

Two-Dimensional Market Heterogeneity

Understanding the anatomy of market strategy begins with the recognition that markets are inherently heterogeneous. But this book goes beyond a call to recognize the heterogeneous nature of markets. I propose the new, more precise and elaborate construct of two-dimensional market heterogeneity as an integrating and organizing principle for market strategy, the backbone of an intellectual anatomy. The two dimensions of this market heterogeneity are product heterogeneity and customer heterogeneity. This structure allows us to separate market strategy into its basic forms.

In behavioral terms, the structural dimension of product heterogeneity refers to the type (or "subject") of a company's behavior. The type of a company's behavior can be divided into two general categories: product behavior or price behavior. The dimension of customer heterogeneity refers to the target (or "object") of a company's behavior. The target of a company's behavior can also be divided into two general categories: behavior directed at all customers or behavior directed at customer segments. The two structural dimensions of heterogeneity thus delineate four fundamental forms of market strategy:

- "Product augmentation": product strategy that appeals to all customers
- "Product segmentation": product strategy that is directed at customer segments
- "Price segmentation": pricing strategy that is directed at customer segments
- "Price competition": pricing strategy that appeals to all customers

The unifying principle (and distinguishing criteria) for all the behavior types is how they exploit two-dimensional market heterogeneity. Each form exploits either one dimension, the other dimension, both, or neither. Product augmentation exploits product heterogeneity. Price segmentation exploits customer heterogeneity (as to price sensitivities). Product segmentation exploits product heterogeneity and customer heterogeneity (as to preferences). And price competition, of course, takes advantage of neither. Both these structural features and the behavioral aspects of each form of market strategy are summarized in Figure 1.1.

Product augmentation is the enhancement of the "product" (in the broadest possible sense of the word) in a way that increases its appeal to all customers. Effective product augmentation reduces the price sensitivity of a firm's customers and diminishes the ability of its customers to substitute its products for competitors' products. In this way it establishes a band of insulation from competition, creates the possibility of receiving premium prices for its product, engenders customer loyalty, and facilitates firm and market stability.

Price segmentation, in its simplest forms, is the varying of the price of a product to exploit differences in customers' price sensitivities. Effective price segmentation extracts higher prices from customers willing to pay more while simultaneously offering lower prices to customers who would otherwise be unwilling to buy at a higher price. In this way, effective price segmentation allows a firm to capture more of the value it adds from the less price-sensitive customer segments while simultaneously extending its penetration of the customer base to more price-sensitive segments it could not otherwise reach.

Product segmentation is the tailoring of products to more closely fit the preferences of a particular segment of customers within the total customer base. All of the same logic that applies to product augmentation applies to product segmentation because product segmentation is segment-based product augmentation. With respect to a particular customer segment, effective product segmentation has the same effect as product augmentation. The difference lies in the fact that product segmen-

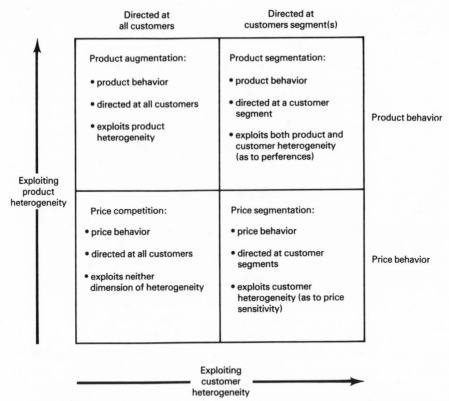

Figure 1.1. Structural and behavioral features of market strategy framework.

tation exploits differences in customers' preferences by targeting a specific segment. It thus combines both the logic of exploiting customer heterogeneity with the logic of exploiting product heterogeneity.

Product segmentation, however, is not simply the use of both product heterogeneity and customer heterogeneity in one type of behavior. Product segmentation is more complex than this. It is the exploitation of one dimension of market heterogeneity by means of the other. Customers are segmented on the basis of a product augmentation. And the product is augmented on the basis of the needs of a customer segment. Product segmentation is thus not only segment-based augmentation but also product-based segmentation.

Product segmentation, price segmentation, and product augmentation imply an emphasis on understanding demand, the customer, and the value he places on a product. Product augmentation attempts to attract customers by improving the net benefits all customers associate with a product, price segmentation entails distinguishing customers on the basis of the value (the reservation price) they place on a product, and product segmentation attempts to add value through a more precise satisfaction of customer needs with augmentations that have a segment-specific appeal. All three of these market behaviors are fundamentally different from price competition, which (1) relies exclusively on reducing a firm's costs per unit of output as a means to competitive advantage, and (2) tends to reduce profitability once cost advantages are exhausted.

Indeed the maintenance of high rates of profitability and continued firm growth is related to the ability to exploit two-dimensional market heterogeneity. The maintenance of extranormal profits depends upon product augmentation, which insulates the firm from the profit-reducing effects of price competition and creates price premiums, and on customer segmentation that, when based on price, leads to the extraction of consumer surplus and, when based on product differences (product segmentation), has the same effect as product augmentation albeit with respect to a smaller target market.

The common language of the demand curve is helpful in further clarifying what each category of behavior means and how it works. In product augmentation, the demand curve is turned or "kinked." In price segmentation, the demand curve is disaggregated or "tapped" at various levels. Product segmentation combines elements from both these sets of logic, while price competition draws on neither. In price competition, the company "rides down the demand curve" to lower and lower price points associated with "normal" profitability. These phraseologies are more fully explained later.

The Economic Logic of Marketing Strategy

The economic logic of demand and the marketing orientation toward the customer are not only complimentary but are flip sides of the same coin. Both are focused on value and neither are focused on costs, as is the case with the three forms of heterogeneous market strategy.

In delineating the three forms of market strategy, this study enhances and extends existing theoretical concepts and meshes disciplinary perspectives, grafting the logic from one area, for instance, onto the objective function of another. Beyond delineating the basic forms of market strategy (each of which individually transcends any particular disciplinary perspective), this study brings these separate forms together in the common principle of exploiting two-dimensional market heterogeneity, thereby unifying previously isolated pockets of analysis in different disciplines into a single framework of market strategy.

"Product augmentation," is based on linking the monopolistic competition logic of "product differentiation" with the "augmented product concept" of marketing. "Price segmentation" is rooted in the economic logic of "price discrimination" but is made more robust by extending it to include marketing techniques that do not entail the use of explicitly discriminatory pricing yet achieve the same effect. The new idea of "product segmentation" borrows from the broader marketing idea of "market segmentation" but is enhanced with fuller yet more precise logic in that it incorporates in one concept both product-dimensional and customer-dimensional logic. A quantum jump in the clarity of understanding each of these three forms is achieved through the alternative frame of reference provided by the nontraditional economic area of spatial competition and, most importantly, the spatial logic of "vertical differentiation" and "horizontal differentiation" (explained in Chapter 2). For each strategy type, Figure 1.2 summarizes (1) the demand logic, (2) roughly

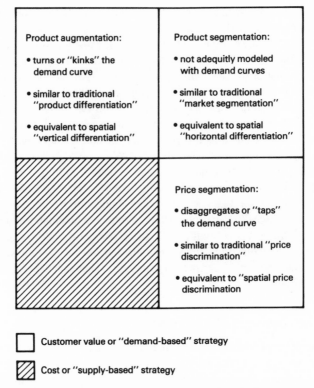

Product augmentation:	Product segmentation:
• turns or "kinks" the demand curve	• not adequitly modeled with demand curves
• similar to traditional "product differentiation"	• similar to traditional "market segmentation"
• equivalent to spatial "vertical differentiation"	• equivalent to spatial "horizontal differentiation"
	Price segmentation:
	• disaggregates or "taps" the demand curve
	• similar to traditional "price discrimination"
	• equivalent to "spatial price discrimination

☐ Customer value or "demand-based" strategy

▨ Cost or "supply-based" strategy

Figure 1.2. Summary of economic logic and equivalent concepts.

similar concepts in traditional economics and marketing, and (3) equivalent concepts in modern spatial economics.

A Metaphor for Market Strategy: Shaping a Cake

The separate areas of economic analysis associated with the different strategy types have been built up and refined through a rich tradition of economic thought. No attempt is made in this book to explicate the mechanics of general equilibrium, price discrimination, monopolistic competition, or spatial competition, although underlying core concepts of these latter three are discussed in the next chapter. However, the logical basis for the different strategy forms can be summarized in a simple way without reference to economic analysis. Such a summary has less depth than the eloquent mathematical and geometric language of economic analysis, but it is sufficient for a rough representation of the logic of the different market behavior types.

A particular product revenue base of a specific firm can be thought of as a three-dimensional cake. The height of the cake is the revenue per customer (the price customers pay). The two dimensional area bounded by the circumference represents the size of the heterogeneous customer base (each location within the circumference can be thought of as representing a different customer with different needs). The higher the cake, the higher the price. The larger the two-dimensional area of the circumference, the greater the size of the customer base (the number of customers). Therefore, the larger the total three-dimensional area of the cake, the greater the firm's total revenue for the particular product market.

Price competition as a strategy entails lowering the height of the cake in attempting to increase its width (horizontal area) or, in other words, increasing the size of the customer base by cutting price. (See Figure 1.3, southwest quadrant). The effect on revenue, of course, depends on price sensitivity, which in turn is related to a variety of other factors. The profitability of such behavior depends on the revenue effect as well as the behavior of costs, which again depends on a variety of other factors.

Product augmentation (northwest quadrant) attempts to increase the size of the cake by increasing its height, increasing its circumference, or both. Effective augmentation allows a firm to charge a higher price at a given volume level, thus increasing the height. Effective augmentation also allows a firm to expand volume for its product at a given price level, thereby increasing the horizontal area. Effective augmentation therefore implies a change in the dimensions of the cake that increases its total area. Which dimensions grow and how much they grow depends

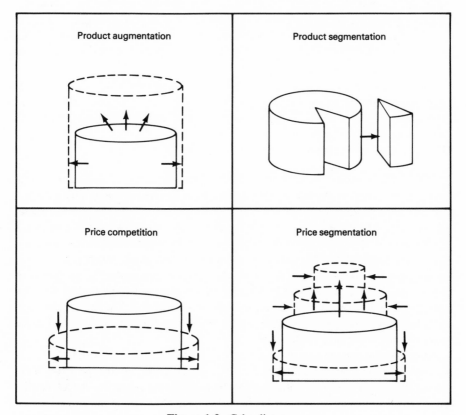

Figure 1.3. Cake diagram.

on pricing and how the firm wishes to trade off price premiums for greater quantity demanded.

Price segmentation (southeast quadrant) attempts to change the simple cake into a cake with layers of multiple heights and widths. Effective price segmentation can entail extracting higher prices for the same product from those customer groups willing to pay more, thereby resulting in higher layers. Effective price segmentation can also entail extracting revenue from customers only willing to pay lower prices, thereby creating wider layers with lower heights. Effective price segmentation thus implies a change in the shape of a simple cake to a more complex shape (such as that of a wedding cake) that increases the cake's total area. The finer the gradations of prices extracted for the same product, the more effective the price segmentation and the greater the total area and, therefore, total revenue.

Product segmentation (northeast quadrant) attempts to carve out a piece of the cake representing a group of customers with common product needs that are distinguished from other sections or locations in the cake's circumferential area.[3] With respect to that segment, or piece of cake, the product segmentor attempts to do the same thing as the product augmenter: increase its height or width or both. Unlike the product augmenter who achieves this through a generic appeal to universal benefits, the product segmentor increases total size by providing a specific augmentation that is needed by only the segment of customers. The ultimate objective of the product segmentor is to offer such a distinctive product to its customer segment that the segment becomes fully insulated from the rest of the market and becomes a distinct market it its own right.

THE IMPORTANCE OF INFORMATION

The Function of Markets

In addition to forming the backbone of the conceptual anatomy described above, the principle of two-dimensional market heterogeneity provides the foundation for a more balanced perspective on the function of markets. In the conventional wisdom of economic theory, the function of the market is the efficient production of firm's outputs and the efficient allocation of those outputs to customers. The key operating variable in this function of productive and allocative efficiency is the price mechanism, which serves to "clear the market." The optimal operation of the price mechanism depends on several conditions, one of the most important of which is product homogeneity. To the extent this condition is violated, the price mechanism operates suboptimally, efficiency is compromised, competitiveness is said to be reduced, and welfare is diminished.

Although the implicit definitions of "competitiveness" and "welfare" in this linkage can be challenged (and they are in the Conclusion), the linkage between the suboptimal operation of the price mechanism and lower efficiency is hard to dispute, and it is not done here. What is challenged, however, is the exclusivity of efficiency in the role of competition and in the function of markets. The introduction of heterogeneity along two dimensions implies the need to balance efficiency with the notion of effectiveness in market behavior.

Efficiency is the relation of input to output. It is a readily quantified construct and is thus often expressed in terms of monetary units, or units of output per unit of time or effort expended. Effectiveness is the relation of output to company goals. As applied to the function of markets, effectiveness can be understood as how well products sold (a firm's output) meet customer needs (a firm's objective). As Bonoma explains: "Marketing effectiveness by definition refers to a comparison of achieved outputs with intended goals. In everyday life we call that comparison [customer] satisfaction."[4]

Without heterogeneity, the function of the market—getting products to customers—merely entails offering products at the lowest possible price. Strategic success lies solely in the ability to gain share through a low cost position and thus superior price competitiveness. Price is the mechanism through which customer wants (demand) and seller's products (supply) are brought together. Hence, in homogeneous markets, efficiency and effectiveness do not conflict.

The function of the heterogeneous market is more complex. If the fundamentally heterogeneous nature of markets is acknowledged and its two-dimensional structure clarified, the view emerges of the function of markets as that of matching: matching heterogeneous products with heterogeneous customers. Matching facilitates the clearing of the market. The more effective the matching, the greater customer satisfaction. The concept of effectiveness in heterogeneous markets therefore implies something beyond what it does in homogeneous markets, for success in getting products to customers is dependent on, in part, success in the process of matching heterogeneous products with heterogeneous customers.

The removal of homogeneity and the introduction of heterogeneity give rise to the potential conflict of market efficiency and market effectiveness. In heterogeneous markets, generating the greatest output at the lowest cost may be detrimental to the matching of different customer wants with differentiated products; it may reduce product heterogeneity important for both general product improvement and the more precise satisfaction of customer wants through segment-specific product behavior. Therefore, while price competition is sufficient for the effective functioning of markets under homogeneous conditions, it is inadequate for the effective functioning of heterogeneous markets. In heterogeneous markets, product and customer dimensional behavior are indispensable to effectiveness.

Two-dimensional market heterogeneity, and its implication for the important role of matching in the function of markets, imply a need to readjust emphasis on price as the exclusive mechanism in the function of markets. Under theoretical homogeneous conditions, all market information collapses into price and, in a literal sense, information is price. Heterogeneity, however, gives rise to the importance of information separate from price and clarifies the relationship of price as a subset of information. Information thus emerges as the superordinate variable in the function of markets, for it is information about heterogeneous products among customers, and information about heterogeneous customers among the sellers of products, that facilitate the market process of matching products to customers required to clear the market.

A more robust analytic apparatus for understanding market behavior begins with the removal of homogeneity as a starting assumption and its replacement with two-dimensional heterogeneity. This results in a more balanced perspective on the function of markets, the objective of that function, and the nature of the mechanism facilitating that function. More specifically, two-dimensional market heterogeneity implies a fuller view of markets to include their function of "matching" heterogeneous products with heterogeneous customers toward the end of not only efficiency, but also customer effectiveness, and through not just the price mechanism, but the superordinate vehicle of market information.

Implications of Information Inadequacy

The significance of market information for market strategy lies in its inadequacy and, in particular, the inadequacy of information about products among customers needed to make purchasing decisions. This inadequacy of market information has specific effects. First, it diminishes the ability of the buyer to select among products (low buyer sophistication) and particularly the ability to reduce products to some "implicit price," or common price-performance measure facilitating product comparisons. Second, inadequate market information increases the customer's receptivity to information signals from the seller that substitute for direct information. Third, and most important, inadequate market information leads to the phenomenon of perceived customer purchase risk. The strategic import of market information rests in the ability of the seller to take advantage of these effects of information inadequacy:[5] low buyer sophistication, high customer receptivity to information signals, and perceived customer purchase risk. All of these effects of inadequate information are intertwined in mutually reinforcing relationships that derive from their common prime source, inadequate information, a phenomenon that is given theoretical rigor by the modern economic concepts of "asymetric information" and "information acquisition" discussed in Chapter 3.

Under perfect information conditions, premium product quality is readily perceived by the customer. Sellers can extract price premiums that reflect the greater utility of higher quality products because customers can reliably assess quality and thus readily compare products on the basis of price. Price behavior is therefore an effective competitive mode and price premiums reflect the actual utility of superior quality and nothing else.

Under imperfect conditions, the seller is unable to extract a premium commensurate with the higher quality level because inadequate information among buyers inhibits their ability to make reliable price/performance or price/quality comparisons among competitive products. As a result, price is less important in determining buying decisions and price competition is less effective in markets with imperfect information. Conversely, product-dimensional behavior that exploits low buyer sophistication, signals information to the customer, or reduces purchase risk that arises from inadequate information becomes even more important than it would be under better information conditions. Thus, not only does information inadequacy diminish the strategic import of price competition but it also intensifies the importance of non-price-oriented market strategy.

"Market signaling" exploits inadequate information by substituting for direct information. The greater the inadequacy of market information, the greater the need for the customer to infer information and evaluate products on the basis of cues or signals, and hence the greater the potential role of cues in market strategy. Perhaps the most important signal to customers is firm "reputation." The importance of reputation arises only as the result of imperfect market information. Under perfect information, reliance on reputation is unnecessary for everything is known about the current product to be purchased. Imperfect information creates the potential for a market strategy of "reputation building," which when successful, can "enable firms to charge premium prices" and create an array of "other potentially favorable consequences."[6]

The phenomenon of perceived customer purchase risk also arises from the inadequacy of market information. Perfect information as to both the outcome and consequences of a purchase decision precludes risk. The importance of customer risk for business strategy lies primarily in the possibility of reducing it as a form of product augmentation and, to a lesser extent, in the possibility of grouping customers according to risk categories. The reduction of perceived risk is a vehicle for exploiting the product dimension of market heterogeneity (and therefore is a form of product augmentation), while the segmentation of customers into groups according to risk types serves as a tool for taking advantage of customer heterogeneity (and therefore is a form of customer segmentation). Further, the segmentation of customers into risk groups can also be a vehicle for price segmentation. In fact, more generally, many forms of price segmentation entail the exploitation of inadequate customer information whether or not risk is involved.

Figure 1.4 highlights how information inadequacy is related to the exploitation of two-dimensional market heterogeneity. Two-dimensional market heterogeneity gives rise to the importance of market information and the potential for exploiting either dimension of heterogeneity, or both, as forms of market strategy. The ascendency of the role of information, which results from markets composed of heterogeneous products and heterogeneous customers, leads to the possibility of exploiting its inadequacy as a further market tool.

Hence conducting market strategy lies in the taking advantage of two-dimensional market heterogeneity and its implication for information. Either dimension of heterogeneity, or both, can be exploited through generic forms of market strategy (product augmentation, product segmentation, or price segmentation) that may or may not entail the exploitation of inadequate information. For example, a reliability enhancement to a product is a behavior that exploits product heterogeneity through an action that appeals to the entire customer base, but it is also a behavior that exploits inadequate information by reducing customer purchase risk that arises from inadequate information.

THE MASTERS OF MARKET STRATEGY:
IBM AND ITS BEST COMPETITORS

The framework delineated above is of practical use to any business: consumer or industrial, large or small, emerging or mature. Indeed the framework has already been effectively applied and used in a variety of businesses. Work at McKinsey &

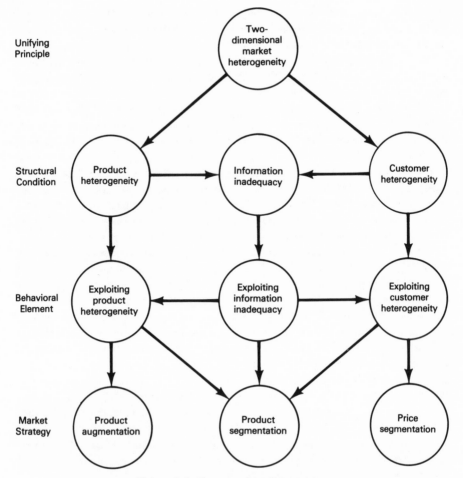

Figure 1.4. Sources of market strategy.

Co. has used the framework to help diagnose, develop, and implement market strategy in the areas of consumer goods, industrial manufacturing, financial services, electronics, and communications media. To take one example, it has been strikingly effective in highlighting new opportunities for segmented pricing by dominant product augmenters in the insurance business resulting from the unique, but unexploited, characteristic of the insurance industry that identical insurance products can legally be explicitly, differentially priced. It also has been useful in academic research. Work by Fondas, for example, found the framework helpful in distilling from firm behavior the essential strategy of business units in five different industries: conduit fittings, hand tools, window treatments, industrial cables, and oil drilling compressors.[7]

Much of the literature on strategy suffers from one extreme or another; it is either too theoretical to be of practical use or it is exclusively descriptive of particular cases providing no generic framework to be applied to different situations. The first part of this book addresses the latter shortcoming by providing generic theory applicable to a variety of situations. The former shortcoming is more difficult to address. Some works that do provide generic frameworks often supplement them with a vast array of briefly sketched, unrelated examples, even one-line examples. But the recitation of shallow and disjointed examples is not particularly instructive to the businessman or student of business seeking to apply the concepts to under-stand and diagnose strategy, or implement the concepts to develop and conduct strategy.

Overcoming this shortcoming requires depth: an intensive, detailed look at in-structive cases of actual market behavior. It is only through such in-depth observa-tion of actual behavior that the robust applicability and practical utility of a frame-work is made clear.

This is accomplished in this book by focusing on a few key competitors in a particular business. The business is computers. And the firms are IBM and two of its best competitors, Cray Research and Tandem Computers. These are masters in the use of market strategy and much can be learned from their product tactics, pricing techniques, and strategic approach. What we find is that the framework helps us understand their behavior and conversely their behavior teaches us how to implement the ideas of the framework. The framework shows how they succeeded and/or why they have failed. In this way it teaches the reader through real, in-depth examples about concepts that are useful to developing and diagnosing strategy in any business. In addition, such in-depth observation reveals some otherwise hidden secrets of IBM and its best competitors, the "tricks of the trade" in implementing market strategy.

Broadening the Benefits Bundle/Revenue through Risk Reduction

The behavior of IBM is an exceptionally good example of product augmentation and the strategic use of the effects of information inadequacy. The concept of product augmentation as developed in this book combines the distilled economic logic of product differentiation with the augmented product concept of marketing. This hybrid concept provides a more robust understanding of IBM's behavior than either the economic or marketing concepts do separately.

The perspective of the augmented product concept, originating in marketing literature, rejects the economic notion of a product as a commodity; it requires that it be understood in terms of its widest potential for providing customer benefits. From this perspective, the product consists of not just the core physical thing itself but the increasingly less tangible layers of customer benefits or potential customer benefits that surround it. And it is precisely in this area of less tangible benefits that significant differentiation or "augmentation" or "product broadening" can create a market advantage.

This is exactly what IBM does. Product augmentation along non-price/performance dimensions, and specifically along less tangible risk-reducing dimen-sions, is a major source of IBM's customer appeal. Empirical economic research, customer survey data, IBM's own internal competitive product assessments, as well as testimony by IBM executives and industry observations by others all suggest that, even in the heyday of its technological innovativeness, IBM's advantage cannot be explained by superior price/performance. Rather it is "nonhardware differences" that account for premium prices IBM is able to extract. In particular, the evidence points towards IBM's image of offering more in terms of "nonhardware services" as a source of market effectiveness.

Moreover, statistical data on purchasing behavior indicate that customers in general purchase mainframe computer products because of customer support, product reliability, and firm stability—factors that appeal to the effects of information inadequacy and, specifically, low buyer sophistication, high receptivity to information signaling, and perceived customer purchase risk. The further statistical fact that those customers who have a greater need for reduced purchase risk are also those customers who select IBM shows that IBM's augmentation strategy to reduce perceived risk does in fact work. As will be detailed later, this strategy is implemented through the marketing techniques of bundling and account management, as well as the extensive use of non–sales field personnel, which both improves the perception of IBM's product as the low-risk choice and intensifies low-risk purchasing behavior.

Surplus through Subtley/Extraction without Elasticities

The role of price segmentation as an important form of market strategy is generally discounted because of legal restrictions on discriminatory pricing and the perceived difficulties of implementing segmented pricing that derive from operational and methodological problems in empirically measuring price elasticities that distinguish customer segments. But the discounting of price segmentation as an important strategic tool betrays a lack of understanding of the subtle, nonexplicit manifestations segmented pricing may take, as well as the fact that several forms can be easily implemented with little customer analysis.

Several types of segmented pricing price-discriminate only indirectly in so far as they do not entail explicitly offering different prices to different customers. Moreover, such forms of segmented pricing do not require the a priori identification of customer groups by determining the differences in price elasticity. Tie-in sales, metering, and product bundling exploit customer heterogeneity, specifically, heterogeneity as to price sensitivities without differential pricing and without elasticity measurement. Bundling, for instance, may facilitate the extraction of "consumer surplus" by a seller without the seller ever knowing the levels and differences in price sensitivities among customers. In fact despite legal proscriptions, methodological problems, and operational difficulties in measuring the price elasticities (assumed to be required for implementation), price segmentation can play a major role in effective market strategy.

IBM is a prime example. IBM effectively reaps the benefits of price segmentation in the absence of any effort to identify customer groups according to elasticity differences. In fact, price segmentation is a key element in IBM's overall strategy. Partially as a result of the phenomenon of customer "lock-in," which IBM encourages, it is able to implement a pricing strategy that extracts higher than normal levels of revenue and profit from its established, less price-sensitive customers at the same time it limits competition among the newer, more price-sensitive customers through unprofitably low pricing (subsidized by established customers).

Perhaps one reason for the failure of the Justice Department to prove its more recent and expensive case against IBM lies in the exceptionally detailed level of observation required before the pattern of segmented pricing emerges. Observation of product and price behavior at the ordinary levels of market analysis—the product line, the product series, or the product model—often reveal no particular pattern of explicit price segmentation. In some cases it is only when IBM's pricing behavior is reviewed at the submodel level that the presence of explicit price segmentation is clear.

But in addition to the subtle implementation of explicitly differential pricing, techniques of price segmentation can be subtle by their very nature in that explicitly different prices need not be charged to exploit price sensitivity differences. IBM's tying of card sales to mainframes, metering usage through extra-use charges, and bundling processors with memory are all ways in which IBM exploits sensitivity differences without charging different prices. These techniques have been modeled in the theoretical literature, but identification of them here shows their practical utility as strategic tools. A careful look at IBM's strategy also reveals other forms of behavior that extract consumer surplus while avoiding explicitly discriminatory pricing (as do tied sales, metering, and bundling) in ways not addressed in the literature and thus not previously considered in formal analyses. These include, the "functional pricing" of submodels, educational discounting, "stable charges" across product generations, interface manipulation, and the variation of product bundles through the disproportionate allocation of field support.

The presence of segmented pricing in IBM's behavior is hard to discern because, with the exception of functional pricing and educational discounting, price differences are not explicit and/or financial accounting methods camouflage the fact that this pricing is not related to cost but is based on exploiting differences in price sensitivity. In the case of functional pricing, as already noted, model-level profit data hide the pattern at the submodel level. In the case of educational discounting, the positioning of this behavior as a charitable action benefiting nonprofit institutions serves to deflect attention from its price-segmenting nature. Irrespective of the legality of IBM's behavior, the evidence shows that such pricing is based not on cost but on the strategic objective of protecting or expanding sales among more price sensitive customers and extracting maximum revenue from less price sensitive customers.

Segmentation without Superficialty/Tenacity despite Temptation

Product segmentation represents the intersection of the two dimensions of heterogeneity. But it is more than simply the exploitation of product heterogeneity and customer heterogeneity in one behavior. Rather, it is based on the more complex logic whereby the exploitation of one type of heterogeneity is a vehicle for exploiting the other type of heterogeneity. It is thus segment-based augmentation and product-based segmentation.

The use of product segmentation requires that conditions be met both on the product side and the customer side. Segments must be defined properly in the first place and the product augmentation must be generated and sustained. More specifically, the effective use of product segmentation in actual markets requires that a segment is selected based on a distinguishing product need and that the firm has the ability to generate, and the willingness to sustain, this segment-specific focus despite the lures associated with a strategy directed toward a broader market.

The common marketing practice of focusing sales efforts on particular locations, particular industries, or particular size categories, although a form of market segmentation, is not product segmentation because it does not segment customers on the basis of product needs. Such demographic, geographic, or industry-based segmentation falls short of the potential of product segmentation because it does not fully exploit both dimensions of market heterogeneity. In particular, it does not focus on the product and, more specifically, the product-augmenting logic component of product segmentation. For instance, geographical location per se says little, if anything, about customer differences as to product augmentations needed.

The concept of product segmentation is not only useful as a tool for developing strategy but also as a device for understanding the reasons for success and failure of a segment-focused strategy in the past. A review of segment-focused behavior suggests that the effectiveness of Cray Research in the super computing niche, as well as Tandem Computers in the fault-tolerant niche, is related to their ability to meet the requirements of product segmentation as to both customer segmentation and product augmentation. The effectiveness of Cray and Tandem lies in the ability to identify segments on the bases of a *product* need in the first place and the ability and willingness to sustain a commitment to a segment-focused policy despite the lures of strategy directed at the general market. Conversely, the ineffectiveness of segment-focused strategies of other firms is due to superficial segmentation in the first place (as is the case with the "BUNCH"[8]), the lack of a clear commitment to a segment-specific strategy from the outset (as is twice the case with IBM), or the failure to sustain the segment-specific focus of a previously effective strategy (as is the case with Control Data Corporation).

Unlike the BUNCH, who failed to define customer segments in terms of product needs, a review of Cray's customer base indicates that Cray targeted a segment of otherwise diverse customers who were defined by an intense, shared, yet distinct, product need. A review of Cray's product actions since its founding reveals a strong commitment and a demonstrated ability to implement segment-focused augmentation. Cray's commitment to its segment-focused augmentation (ultrafast computing) is not only evidenced by its ongoing record of increasing the limits of computing speed, but also in a demonstrated willingness to sacrifice product reliability, compatibility, and general purpose utility in order to achieve greater speed. Cray's commitment is also apparent in its rejection of product-line broadening projects. Quantifiable evidence of Cray's commitment can be found in the biased distribution of research and development (R&D) and marketing resources toward Cray's traditional customers, despite the potential for greater revenue from customers with more IBM-like need profiles.

On the other hand, the repeated ineffectiveness of IBM's segment-focused strategies is related to a lack of commitment and ability to implement segment-focused augmentation. Similarly, the ineffectiveness of CDC is related to its failure to sustain adequate focus and specifically its failure to resist the enticement of greater revenue and reduced strategic risk associated with market-broadening behavior. CDC's case, in particular, provides a clear example of the danger of shifting between, or mixing, a segment-focused strategy with a broad market appeal strategy. Such behavior invites competition at both the high end and low end of a firm's market. This is precisely what happened to CDC, who was squeezed between IBM's superior general purpose computers at one end and Cray's faster supercomputers at the other end.

SIX BASIC RULES FOR ENHANCING STRATEGY

For each form of heterogeneous market strategy summarized in the first part of this chapter (product augmentation, price segmentation, and product segmentation), a few important lessons emerge from the detailed, in-depth study of the masters of market strategy just summarized above. There are six generic lessons or guidelines, two for each of the three types of heterogeneous market strategy:

- Product augmentation
    - —broadening the benefits bundle
    - —revenue from risk reduction
- Product segmentation
    - —segmentation without superficiality
    - —tenacity despite temptation
- Price segmentation
    - —extraction without elasticity
    - —surplus through subtlety

I have found these lessons helpful in developing, implementing, and diagnosing market strategy in a variety of businesses. An effective mechanism for remembering these guidelines is to reorder the six of them to form a nmenonic; this nmemonic coupled with the alliteration of each phrase is a useful memory trick (see Figure 1.5).

ORGANIZATION OF THIS BOOK

This book is structured in much the same way as this chapter: it is divided into two parts (excluding chapters 1 and 7, the Introduction and Conclusion). The first part is theoretical and generic; the second part is applied and specific. The basic concepts for the market strategy framework have been sketched out in this introductory chapter. Readers less interested in theory may wish to skip to Part II, the applied section; but those interested in a deeper understanding of the economic and marketing logic underlying the new idea of exploiting two-dimensional market heterogeneity should continue without skipping. This first section is in no way a literature

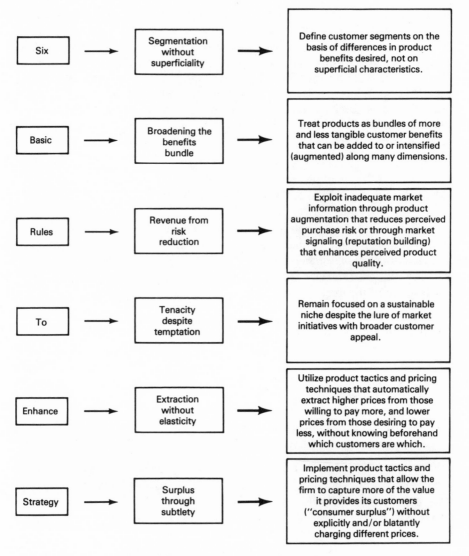

Figure 1.5. Six basic rules to enhance strategy.

review but the integration and enhancement of multiple perspectives to create a new framework.

This first part consists of two chapters (2 and 3). Chapter 2 focuses on the conceptual components of two-dimensional market heterogeneity:

- Exploitation of product heterogeneity and product augmentation
- Exploitation of customer heterogeneity and price segmentation
- Exploitation of both dimensions of heterogeneity and product segmentation

It draws on the logic not only of the traditional economic and marketing areas of product differentiation, price discrimination, and market segmentation, but also the modern area of spatial competition and its concepts of vertical differentiation, horizontal differentiation, and spatial price discrimination.

The next chapter in Part I (Chapter 3) focuses on the implications of two-dimensional market heterogeneity:

- Importance of information in the functioning of markets
- Potential for exploiting its inadequacy as a source of market strategy

This chapter does not draw on traditional economic and marketing concepts but new areas in economics, such as "information acquisition" and "asymmetric information," and areas in psychology and marketing, such as information signaling and customer risk.

Part II of this book is much different. It takes the framework developed in Part I and applies it to IBM, Cray, and, to a lesser extent, Tandem, Control Data Corporation, and the BUNCH. In this applied section there are three chapters. The first (Chapter 4) shows how IBM uses product augmentation and customers' risk reduction as a key component of its market strategy. The next chapter (Chapter 5) shows how IBM cunningly implements a market strategy of price segmentation. The last chapter (Chapter 6) shows how some competitors have successfully circumvented IBM through a strategy of product segmentation (Cray and Tandem), why others have failed in niche strategies, and also why IBM has failed to respond effectively to the success of its best niche competitors.

NOTES

1. G. Stalk, T. M. Hout, *Competing Against Time, How Time-Based Competition is Reshaping Global Markets* (New York, Free Press, 1990), p. 105.

2. While the competitive strategy area has taken major strides in developing tools for thinking about business behavior it has provided only shadowy illumination of the customers' side of strategy because, as Michel Robert argues, it focuses on competitive rivalry (and the forces affecting this), only one aspect of the environment to which a firm must relate: "Corporate America has never been as engrossed in techniques and formulas of competitive analysis and strategy as in the last ten to fifteen years. . . . These approaches to corporate strategy are flawed, and for one reason. Their whole thrust is based on the assumption that corporate strategy starts with an analysis of competitive position. This assumption is a very myopic view of strategy." Such "myopia" is understandable, argue Reed and De Fillippi, because the source of much of competitive strategy thinking, industrial organization economics, focuses on competition within industries: "The original intent of industrial organization thinking was the maintenance of intra-industry competition. This intent has been inverted to produce models aimed at helping firms realize supernormal returns by circumventing competition. Given the IO roots of much strategy thinking, it is not surprising that establishing and sustaining a competitive advantage has been given so much emphasis." References for quotations in the order cited are as follows: M. M. Robert, "Managing Your Competitor's Strategy," *Journal of Business Strategy* (March-April, 1990), p. 24; R. Reed and R. J. DeFillippi, "Causal Ambiguity, Barriers to Imitation, and Sustainable Competitive Advantage," *Academy of Management Review*, 33, No. 2 (June, 1990), p. 233.

3. Note in this illustration the piece of cake, not the whole cake represents the firm's price level, the size of the firm's customer base and thus the firm's total product revenue; the whole cake represents the total market.

4. T. Bonoma, *The Marketing Edge: Making Strategies Work* (New York, Free Press, 1985), pp. 191–192.

5. Note that the strategic import of information inadequacy also lies in the ability of a firm to

contribute to its cause through, for instance, frequent product changes that, as noted later, tend to exacerbate inadequate information conditions.

6. C. Fombrun, M. Shanley, "What's in a Name? Reputation Building and Corporate Strategy," *Academy of Management Journal,* 33, No. 2 (June, 1990), p. 233.

7. N. J. Fondas, "Managerial Agendas and Strategic Alignment: A Field Study of General Manager Behavior" (Harvard Business School doctoral dissertation, 1987).

8. The "BUNCH" is an acronym for Burroughs, Univac, National Cash Register, Control Data, and Honeywell, three of which no longer exist as such due to mergers and acquisitions.

I
The Idea of Two-Dimensional Market Heterogeneity

2

The Conceptual Components
of Two-Dimensional Market Heterogeneity

> Economic analysis typically was not and could not be integrated within the taxonomic, eclectic framework of marketing literature. Consequently, economists seeking illumination from marketing literature for the most part found little or nothing that would relate to the accepted categories and theories of competition [and] marketing teachers and scholars increasingly became cut adrift from economic theory, and hence became disinterested in its use, refinement, or creative adaptation and development. . . . The literature and results both in economics and in marketing might have been vastly different if this type of differentiation had not occurred.
>
> E. T. GRETHER[1]

> If one approaches economics expecting too much, one may well come away with too little. Economic models are not designed to . . . provide practical algorithms for implementing [marketing] strategies. But marketing academicians and practitioners, [who] are soon disillusioned if they look to economics for practical solutions to [specific] problems, would be shortsighted to label the theoretical models of economics irrelevant. . . . Economic models may be weak in specific prescriptions for individual action, but they are strong in useful heuristics for understanding.
>
> THOMAS NAGLE[2]

PRODUCT HETEROGENEITY
AND PRODUCT-DIMENSIONAL BEHAVIOR

Product Differentiation

Product differentiation pervades actual market behavior. Yet despite its conspicuousness, its impact in altering theory and in spurring empirical investigations of its causes has been slow and limited. Caves and Williamson point out that "empirical research . . . has failed to identify the empirical basis of product differentiation."[3] Samuelson suggests that the monopolistic competition framework was not incorporated into mainstream economics until recently.[4] Atwater argues that the framework was quickly submerged beneath other contributions and proclivities of the economics discipline: "the changes Chamberlin brought about in economics were overshadowed for decades in the economic literature by the Keynesian revolution and economics' emphasis on mathematical models."[5] Peterson argues that even the one branch of theory that has most fully incorporated differentiating behavior into its framework—IO economics—has done so in a tautological manner that results in "implicit theorizing [that] reduces the credibility of industrial organization analysis in teaching, research and public policy."[6]

The Economic Logic of Differentiating Behavior

The importance of the concept of product differentiation to this study rests in the basic economic logic underlying it. This logic is rooted in the broad model of monopolistic competition first sketched by Chamberlin.[7] The fundamental contribution of Chamberlin's work to this study is his identification of the importance of product differentiation as a means to gain some degree of influence over the price of a firm's product offering. This ability to control price is the variable distinguishing the product differentiator from the pure price competitor. Chamberlin explains that "a sole prerequisite to pure competition is . . . that no one has any degree of such control." From this singular basis for distinction from pure competition, flow the two fundamental assumptions of the model: "there must be a large number of buyers and sellers" and "goods must be perfectly homogeneous," so that no one can exercise any "degree of control over the price."[8]

The basis of Chamberlin's work lies in the recognition that the second assumption is not generally valid. The pervasive violation of this second assumption in a market of many sellers leads him to define a new mode of behavior: monopolistic competition. The central feature of monopolistic competition is, in structural terms, product heterogeneity and, in behavioral terms, product differentiation.

In theoretical terms, the "control over price" that effective product differentiation confers means that a firm's customers are, to some degree and within some range, insensitive to price. This allows the firm to profitably charge premium prices either as a result of increasing price over competition or not responding to competitors' price cuts: opportunities unavailable to the pure price competitor to whom "price is a parameter."[9] More technically, it implies that the horizontal demand, and therefore perfect price elasticity and cross–price elasticity, give way to a market condition of diminished customer price sensitivity and diminished product substitutability. In graphical terms, product differentiation "twists" or "kinks" the individual demand curve facing the firm.

Product differentiation does not change demand by merely changing the independent variables of the demand function, which in turn affect the dependent variable (quantity demanded); rather, product differentiation changes the very nature of the relationship between the dependent and independent variables so that a change in an independent variable (price) will affect the dependent variable (quantity) in a different way. This point is fundamental but does not always seem clearly recognized in marketing literature. For instance, Dickson and Ginter distinguish between a "product differentiation strategy" and a "demand function modification strategy."[10] They argue that "the distinction between product differentiation and demand function modification must be clearly recognized and understood." Yet monopolistic competition theory shows how product differentiation that changes the product, or the perception of the product, can change the demand function itself and, more specifically, the functional relationships between price and quantity demanded (price elasticity).

The distinction these authors seem to be making, although seemingly mislabeled, is actually the important distinction between two variations of the same product differentiation strategy: changing products to fit customer wants versus changing customer wants to fit existing product differences. But contrary to what these authors seem to imply, both types of behavior rest on the economic logic of altering the functional relationship between price and quantity demanded. The recognition that both these types of behavior are flipsides of the same coin is clarified as a result of a major theoretical implication of this study that, along with this distinction, is discussed in Chapter 3 (in the section "From Market Price to Market Information").

The significance of product differentiation for business strategy lies in its potential to protect a firm from the profit reducing threat of price competition or, in other words, "escape, at least in part, the rigid control exercised over undifferentiated goods by market forces."[11] In terms of competitive strategy, product differentiation creates a protective band of insulation (in the form of reduced customer price sensitivity and product substitutability) surrounding the differentiating firm that reduces competitive threats, engenders customer loyalty, and may contribute to firm stability if not the stability of the entire market.

A Limited Perspective on Product Dimensional Behavior

Understanding the shortcomings of the monopolistic competition model for purposes of this study requires that the price competition model be set in the context of selected aspects of intellectual history and historical fact. The price competition model can be viewed in at least two ways: as an essentially accurate representation of actual behavior and as theoretical ideal point toward which policy might strive. The reason for the attention given by early economists (before Marshall) to the price competition framework arguably is rooted in the first way of viewing the model, as a relatively accurate generalized description of reality.

The nature of economic activity reinforced by the political climate of that time support this view. Market activity, involving the production of inherently homogeneous agricultural goods and the mass production of standardized products in the early industrial revolution, set the stage for a competitive environment in which many firms competed on the basis of price in commodity or near-commodity markets. The emphasis on the role of price by economists was thus "indicative of an excess demand for a largely undifferentiated supply of goods and services such that price was the single most important determinant of how that supply would be allocated."[12]

The political climate reinforced this economic reality. Centralized government increasingly gave way to an emphasis on individual liberty; western economies increasingly relied on the role of private actors in economic affairs. Economic activity became more decentralized, and the price system emerged as the mechanism for coordinating the independent activities of businessmen.[13]

Because the price mechanism facilitated the coordination of decentralized economic activities in undifferentiated product markets without requiring government involvement, it emerged as a key link in the compatibility of a political philosophy with economic progress. The heavy reliance on the price system naturally attracted the attention of economic thinkers. Any preoccupation with the price competition model can thus be understood as logical reaction to the great importance of a working mechanism of actual market behavior.

Eventually, the increasing output of homogeneous goods contributed to the increasing satiation of commodity demand, while higher income levels contributed to the dissatisfaction with product homogeneity and the desire for heterogeneous goods more precisely fitted to variable consumer tastes. These phenomena, coupled with increasing industrial concentration, diminished the empirical, descriptive accuracy of the price competition model.

The diminution of the empirical bases for the price model coincided with the trend in academic thought that removed the study of the price system further and further away from factual institutional detail and actual observed behavior. Economic thinking ascended to new heights of theoretical elegance with the development of general equilibrium theory, which was far removed from the institutional richness and observation-based postulates of Smith's work a century earlier. Although this model of general equilibrium was first presented by Walras only a century after Smith, Samuelson explains that, "it was not until after World War II that economists generally began to think in terms of general equilibrium."[14]

The mutually reinforcing directions of intellectual and economic history set the stage for a reduced concern for closely linking theory to observation and, arguably, the failure to fully appreciate the relative diminution of the role of price behavior in the actual behavior of markets increasingly characterized by industrial concentration and product differentiation. Indeed, consistent with the position of this thesis, Samuelson further argues that even though the recent adoption of general equilibrium analysis "represents an advance in logical clarity" it also represents "something of a retreat in terms of realistic appraisal of actual . . . competitive market structures."[15]

The study of the price system, protected from both logical challenges and tests of empirical validity through more and more narrowing assumptions, led to far-reaching conclusions. It came to demonstrate that price-competitive behavior maximized efficiency and optimized consumer welfare, and thus it came to view pure price competition as the ideal mode of market behavior toward which efforts of public policy are, to this day, directed. While modern views of competition and approaches to competition policy acknowledge the lack of feasibility in many markets of pure forms of price competition, the view that price competition is the "first-best," albeit theoretical, case of market behavior is retained; even the most preferred feasible state of actual market behavior is viewed, at most, as "second best." It is here suggested that the reason for attention to the price-competition model, while still rooted in its immense utility in coordinating economic activity, has shifted relatively with time from its empirically based explanatory power to its theoretical implications for productive efficiency and consumer welfare.

Linked to the mindset that places greater relative emphasis than previously on the theoretic implications and hence normative prescriptive aspect of the price model (as opposed to the explanatory power and hence positive descriptive features of the model) is a perspective that views market behavior in terms of a unidimensional continuum. The theoretical best case (pure price competition) forms one endpoint of this continuum and the theoretical worst case (pure monopoly) is at the other extreme. This is precisely the cast into which Chamberlin poured the substance of the monopolistic competition model of behavior: "regard monopoly as the antithesis of competition . . . at the other extreme is pure competition . . . between the two extremes there are all gradations [in which] both elements are always present."[16]

Because the monopolistic competition model began with a view that constrained market behavior to the narrow line forming the continuum between pure price competition and monopoly, the monopolistic competition model of behavior logically viewed differentiating behavior, which reduced the level of price competition, as a step along the line toward monopoly; differentiation was not another type of competition, no less competitive than price competition, but rather a mixture of competition and its antithesis, monopoly. The product differentiator did not meet the ideal of the *price* competitor; he was a *monopolistic* competitor. His behavior was "monopolistic"; it interfered with the price mechanism of "pure" competition and thus reduced the level of competition, productive efficiency, and welfare.[17]

Consistent with this perspective, product differentiation was viewed as primarily an artificial creation of the seller rather than an inherent potential in the structure of markets; it was pushed on the customer by the seller's desire to gain monopoly power rather than pulled by the heterogeneous nature of customer wants. As prominent marketing scholar and practitioner Alderson explains, the dimensionality Chamberlin added to products was not carried over to customers: Chamberlin did ₁ot recognize "that markets are radically heterogeneous on both the supply side *and the demand side*"[18] [my underscore]. In an extreme interpretation of Chamberlin, "the consumer was the producer's dupe, fooled by advertising into paying low but excessive prices."[19] The author of this statement concludes that "the principle shortcoming of Chamberlin's 'monopolistic competition,' [is] the notion that it was only the sellers' desires for monopoly power that led them to differentiate their products, an implicit carryover from pure competition/pure monopoly modes of thought."

The term "monopolistic competition" itself is revealing of this underlying mindset. The label is used to describe market behavior that is not purely competitive because it is mixed with monopoly elements that pollute (make "impure") the competitive ideal. But what Chamberlin calls "elements of monopoly" or "monopolistic behavior" is, in fact, product differentiation: a behavioral characteristic, not of a monopolist, but of many firms producing heterogeneous products. The behavior of product differentiators is seen to contain elements of monopoly, not because the behavior is the same as that of monopolist, or even that any small feature of the behavior is the same, but because of the singular limited similarity of the effect of product differentiation behavior to the effect of structural monopoly: freeing the firm from the necessity of being a "price taker."

Chamberlin's basis for connecting monopoly elements with product differentiation is hence neither structural nor clearly behavioral. In no way does product differentiation imply monopoly in the true structural sense of the term (one seller). Nor does it necessarily imply any structural condition that is closer to monopoly than the structural conditions for price competition, as is the case with oligopoly (a small number of large sellers). In fact, product differentiation can exist at any level of industrial concentration except one—monopoly structure.

Similarly, in behavioral terms, there is no clear link. Product-differentiation behavior does not make sense for the monopolist; it is redundant; indeed, it is definitionally impossible. Ironically, in terms of behavior along product dimensions (non–price dimensions), monopoly is closer to pure price competition than product differentiation. Monopoly, like price competition, implies product homogeneity, whereas the distinguishing feature of product differentiation is the exploitation of the potential for differences in the products offered by competing firms.

The model of pure monopoly is misleading in the help it provides in understanding product differentiation. While the similarity of differentiation and monopoly power in the control over price that both behaviors facilitate is clear, further metaphor is unhelpful and positively detrimental to clarity. One of the foremost marketing theorists in the history of the field, Wroe Alderson, argues that "such terms as pure competition and pure monopoly have little relevance [for understanding] competition among business firms."[20] Alderson suggests that the preexisting unidimensional mindset that caused "'pure competition' [to be] taken as a point of departure" upon which monopolistic competition theory was developed, limited Chamberlin's contribution: "If Chamberlin had been the first major student of the subject, he might have moved more directly toward the creation of an appropriate theory."

It is here suggested that the same incomplete unidimensional assumptive foundation that limited Chamberlin's contribution also inexorably inhibited mainstream economists in the possible development of a more radical model of behavior. The use of price competition as an ideal type[21] and the myopic view of all behavior in terms of a linear range between the ideal and the worst cases explain the tenacious focus on price competition as the exclusive measure of competition and the tendency to characterize certain non–price behavior as "anticompetitive" and detrimental to welfare. Laying the groundwork for a more complete and balanced view of market behavior requires a readjustment of perspective on differentiating behavior. The implications of this readjustment for public policy and ideas on welfare are discussed in the Conclusion (Chapter 7).

Product Augmentation

A Narrow View of Product Differentiation

It is recognized among scholars within the marketing discipline that the predominance of literature in the subfield of product policy focuses on consumer markets.[22] Examples used in economic works illustrating product differentiation similarly focus on consumer goods. Both fields typically treat industrial markets as commodity markets. Given this consumer goods focus, both areas place emphasis on the role of what economics terms "advertising effects" and "cosmetic changes" or what marketing writers describe generally as "promotional activity" in differentiating behavior.[23]

The implicit assumption behind this emphasis on consumer products is the notion that the role of differentiating behavior is much less significant for behavior in industrial product markets than in consumer goods markets. The contention that advertising techniques; superficial, cosmetic product differences; and other forms of promotional activity play a larger role in marketing to consumers than in marketing to industrial buyers is not debated here. What is argued, however, is that even if the contention above is correct, one cannot therefore conclude that differentiating behavior is of less significance in industrial markets. Such an implication results from equating differentiating behavior with promotional activity.

In this study, differentiating behavior is viewed as a far broader concept than promotional activity, which is treated as one of many subcategories of product differentiation. The perspective that views differentiation primarily in terms of consumer markets and promotional behavior is symptomatic of a myopic view of product differentiation and ultimately the very concept of a product. The significance of product differentiation for industrial marketing policy does not lie in advertising or promotional techniques; rather, it lies in what is termed in this study "product augmentation"—a term borrowed from marketing literature but redefined and extended here.

The Augmented Product Concept

The "augmented product concept" was formulated and popularized by Levitt, and it is from Levitt that the concept finds it fullest expression. Several themes are discernible in the limited literature on augmentation. Basically they are all linked by the simple proposition that a product is not a physical thing: a product is always differentiable; a product is a bundle of consumer benefits; a product is a solution to a customer problem; a product is a system.

Levitt portrays this definition of a product diagramatically with his representation of a product as a series of concentric rings. The innermost ring reflects only the most limited understanding of a product: the "core product" or the commodity product. Outer rings represent increasingly less tangible product benefits such as service, reputation, and company image. The product augmentation concept calls for a product to be viewed in terms of its widest benefit potential. It thus begins with the recognition that "there is no such thing as a commodity . . . this is even more so for industrial markets."[24] A product is "more than the generic thing produced in the factory." It is a "complex cluster of value satisfactions." Similarly Corey writes:

the "product" is what the product does; it is the total package of benefits the customer receives when he buys. . . . Even though a product might, in the most narrow sense, be indifferentiable, an individual supplier may differentiate his product from competitive offerings through service, product availability, and brand image.[25]

The most recent expression of the essential features of the augmented product concept is provided by Cardoza:

What is a Product? Products or services may be defined by their tangible, physical attributes, or by the specific operations performed in a particular service. From a marketing viewpoint, products or services may be more usefully defined in terms of the benefits they provide, especially as those benefits are perceived by the consumers for whom the products or service is intended. Charles Revlon's epigram sharply points up the distinction between physical properties and perceived benefits: "In the factory we make cosmetics. In the store we sell hope" . . . Manufacturers augment physical products with services such as software for computers. . . . In these instances, the product-with-services combination is often a system.[26]

The importance of this concept lies in the fundamental recognition that "the product" in the eyes of the consumer is more than just the core physical characteristics and technology it embodies. A computer, for example, should be viewed not simply as a piece of hardware but as a cluster of benefits. These benefits not only include basic product performance and functionality, but the product's reliability, speed of delivery, technical support, maintenance service, applications support, product reputation (and thus reduced buyer risk), as well as the seller's help in solving diverse customer problems.

Extending the concept of product augmentation so as to include *net* benefits, and not simply benefits, brings to the fore the importance of addressing opportunities to reduce customer cost as a component of effective market strategy. This refinement calls for a perspective on a product's price as only one component of cost to the customer. Customer costs also include, for instance, installation costs, delivery costs, the costs associated with the details of the contractual financing and purchasing arrangements,[27] conversion costs, and the risk of product inadequacy or failure.

The actual perception of products by consumers as a cluster of benefits and costs is supported in the literature on the psychology of buyer decision making. Bonoma and Shapiro thus point out that: "research into the psychology of decisions provides . . . findings useful to understanding buying motives: buyers act as if a complex product or service were decomposable into bundles of benefits."[28] Yet the assumptions required for "ideal" market behavior from the economic perspective preclude such an augmented product notion. In the ideal case, the economist's product is the "thing itself": a commodity. Similarly, the augmented product concept notwithstanding, "the literature in marketing tends to focus on the physical product."[29] Levitt argues that "production minded executives and most economic theorists are resolutely attached to the idea that goods have intrinsic properties . . . rather than conveying benefits to buyers."[30]

The recognition that a product is a cluster of value satisfactions that include but extend beyond its core, leads to a marketing approach that views a product as a solution to a problem, and purchasing activities as problem-solving activities. Effective marketing policy thus rests, not in selling physical commodities, but in solving consumers' problems. Relatedly, "customers attach value to products in proportion to the perceived ability of these products to help solve their problems."[31] A marketing approach of "selling solutions" can be especially effective when it takes the form of strategies that reduce the customers' perceived risk and strategies sensitive to customer sophistication levels. These concepts are developed in the second half of Chapter 3.

Linking the Economic Logic of Product Differentiation with the Marketing Concept of Product Augmentation

The idea of the augmented product belongs exclusively to the realm of marketing. The framework for the concept does not incorporate the perspective of economics and particularly the economic logic of product differentiation. Indeed marketing literature in general has not built conceptual bridges to the economic logic identified in Chamberlin's monopolistic competition theory. After a vast and comprehensive review of marketing literature, Grether concludes that Chamberlin's ideas have had no general or systematic impact on marketing thought.[32] He finds that only two, out of thousands of marketing texts, incorporate Chamberlin's monopolistic competition concepts into their work. Indeed the evolution of marketing thought as portrayed in history of marketing ideas literature generally gives no recognition to Chamberlin.[33] E. T. Grether writes: "The only systematic endeavor to portray the history and development of marketing thought makes no specific reference to Chamberlin. Bartels, in twelve chapters and in a lengthy bibliography of marketing literature containing many hundreds of publications and writers, somehow overlooked both Chamberlin and his classic volume."[34]

This thesis takes the marketing concept of product augmentation and incorporates it into a larger market behavior framework rooted in the economic analysis of customer demand. The association of the economic logic of differentiation with the marketing concept of product augmentation implies a more precise and analytically rigorous definition for the marketing term than that previously ascribed to it. Once set in the light of demand analysis, the potential power of augmentation for creating high levels of profitability and insulation from the effects of price competition is clarified. Similarly the economic concept of differentiation is improved by adding the marketing perspective of product augmentation. Because the augmented product concept requires the product to be viewed in terms of its widest potential for providing customer benefits, the addition of this concept to the idea of product differentiation results in a concept of product dimensional behavior not as a "monopolistic" behavior necessarily detrimental to consumer welfare, but as an alternative form of competitive behavior essential to customer satisfaction. "Product augmentation," as developed in this thesis, is thus a customer benefit-centered market strategy that provides an operational approach to implementing the abstract economic logic of twisting the demand curve so as to reduce product substitutability, engender customer loyalty, and protect the firm from competition.

CUSTOMER HETEROGENEITY AND CUSTOMER-DIMENSIONAL BEHAVIOR

The Economic Logic of Price Discrimination

Customer heterogeneity, as distinguished in this book from product heterogeneity, refers to differences in customers (not products). It is a precondition to "customer segmentability," which is explained later. Since this section discusses only customer heterogeneity, it is sufficient here to point out that in the absence of product heterogeneity the basis for customer segmentation lies in the product's price. This study uses the term "price segmentation" to refer to customer segmentation on the basis of price.

The economic analysis of price segmentation is rooted in marginal utility theory, which provides a theoretical rationale for the demand curve.[35] The demand curve is a powerful heuristic for explaining the logic of price segmentation. Long points out that in the economic framework of demand analysis "consumer demand . . . is represented by a complete set of consumer reservation prices."[36] The reservation price is that "price just low enough to overcome the consumer's reservation about purchasing an extra unit . . . it is an index of the value of an extra unit of consumption."[37] The basis of price discrimination lies in the ability to exploit the heterogeneity of customers with respect to reservation prices.

In essence, the analysis rests on the notion of marginal utility. The reservation price a consumer ascribes to a good varies inversely with the quantity because the utility of each additional good—the utility "at the margin"—diminishes as the quantity purchased increases. From this starting point the marginalist school, in the early development of economic theory, posited the equimarginal principle whereby a consumer purchases quantities of a good up to the point where the marginal utility derived from the last dollar's worth of each good purchased equals that of all others.

The differences in prices of price–quantity combinations that consumers are willing to purchase creates the potential for various forms of price segmentation. The idea of price segmentation is to receive payments from consumers that vary in such a way as to take advantage of differences in what consumers are willing to pay. In this way price segmentation can generate greater revenue and profits than a price policy that does not exploit this aspect of customer heterogeneity: higher prices to those willing to pay more can increase margins and thus revenue and profits; lower prices to those willing to pay less can expand volume and thus also increase revenue and profits. In the terminology of Marshallian economic analysis, price segmentation aims at extracting the highest possible level of "consumer's surplus":[38] the surplus value or utility to the consumer, not paid for by the consumer, represented by the total area under the demand curve over the price line. Effective price segmentation hence interferes with the natural forces of price competition, protects against the profit minimizing impact of price competition, and results in profits that may in fact exceed those associated with monopoly pricing behavior.

Pigouvian "third-degree" price discrimination is an especially important concept of this study because, unlike the ideal of "first-degree" discrimination and Pigouvian "second-degree" discrimination, third-degree discrimination embodies the notion of disaggregation of demand. Pigou showed that by dividing a customer market into sufficiently separated submarkets with different price elasticities, more revenue and profits could be extracted by charging different prices for the same product than by charging these separate markets the same price.[39] Extraction of consumer surplus is maximized by charging different prices in each submarket so as to equate marginal revenues of the submarkets. The contribution of the notion of third-degree discrimination for this study lies in the identification that the segmentation of customer demand into subschedules of demand can give rise to profits in excess of those resulting from a purely price-competitive behavior mode.

Marketing literature suggests that the price-discrimination framework has also contributed to the identification of the importance of customer heterogeneity in that it set the groundwork for the establishment of an explicit criterion for optimal segmentation definition. Marketing literature often cites Robinson as the source of this criterion because a 1956 article in the *Journal of Marketing* (discussed in the next section), which first recognized the marketing significance of this economic framework, drew on Robinson's work only (although as noted previously the significance of the subdivision of demand concept was established earlier than this in economic thought). Robinson clearly sets out this criteria when she explains:

Total demand [can be divided] into two classes such that the highest elasticity of demand in one class is less than the least elasticity of demand in the other class . . . the parts [can] again be sub-divided and so forth until the point is reached at which each sub-market consists of a single buyer or group of buyers whose elasticities of demand are the same.[40]

Assail points out that the economic "criterion for maximization can easily be translated in clustering terms to mean a minimization of within group variance in elasticities and a maximization of between group variance."[41] In other words, if elasticity can be determined on the individual level, then customer segments can be defined on the basis of similarities and differences in elasticities. The problem of optimal segmentation structure is thus subject to analysis through multivariate statistical techniques and particularly cluster analysis, given the empirical determination of individual price sensitivities.

Price Segmentation: Beyond Price Discrimination

Given legal prohibitions to price discrimination as well as the practical difficulty of defining, a priori, customer groups with distinctly different elasticities, efforts to extract income from customers according to how they value a product, rather than what the product costs, have taken subtle behavior forms not immediately discernible as methods of price discrimination. As Sharpe explains:

The antitrust laws make explicit price discrimination an exceedingly hazardous practice; [sellers] often attempt to obtain the same effect in more subtle ways. In particular they may engage in multipart pricing, offering the same terms to all, even though the actual amounts paid by customers will vary in ways not entirely related to costs. Such practices are also illegal, but their use is far more difficult to prove than outright discrimination.[42]

Such subtle techniques of extracting consumers' surplus without explicit discriminatory pricing are included in the broader category of price segmentation.

In sharp contrast to the current dearth of attention to extension of the monopolistic competition framework, the topic of price segmentation is associated with a fairly robust modern literature. Price discrimination is distinguished from other forms of price segmentation in part because it involves only one product. Other types of price segmentation are just now forming an area of economics that has much to offer the marketing field. Tie-in sales, product metering, and product bundling, for instance, all segment customers and extract a surplus through different techniques that involve more than one product and that do not involve the explicit charging of different prices.

Price Segmentation through Metering

Segmenting customers and extracting different price levels for the same product on the basis of use intensity through a metering device is a subtle technique for price segmentation. The underlying logic as to why usage metering can serve as a tool of price segmentation lies in the understanding that consumers generally place a higher value on a product the more intensely they use it. The intensity of usage as measured through metering serves as a crude proxy variable for reservation prices. Moreover, it also serves as the means for implementing the price segmentation scheme. By charging customers not by the unit of the product, but according to use, the firm is segmenting customers according to use intensity. In effect, the firm is charging different prices according to how valuable the product is to the consumer and thus extracting part of the consumer's surplus.

Segmentation through metering is distinguished from pure price discrimination by the fact that it requires a metering device. It is, however, readily understood in terms of price discrimination as it involves the sale of only one product. It is the redefinition of the "product" from a physical unit, to some unit of usage, that creates the possibility of segmentation. (Chapter 5 reveals how IBM's practice of "extra-use" charging is, in fact, a metering-based price segmentation strategy.)

Price Segmentation through Tie-In Sales

The economic logic of tie-in sales is the same as that just described for metering. The distinction lies in the fact that the tied good is, in effect, the metering device. Price segmentation in this case is implemented by "tying" the sale of one good (the asset) to the sale of another good (the monitoring device) vis-à-vis the terms and conditions of the contractual purchasing arrangement.

In the United States, the Clayton Anti-trust Act prohibited the tie-in sale with the rationale that tying arrangements were an attempt to extend monopoly power from one product to another product: "the illegality in tying arrangements is the wielding of monopolistic leverage; a seller exploits his dominant position in one market to expand his empire into the next."[43] This legal argument suggests that tying allows the seller to gain a dominant market share in a competitive product market by linking the sale of the competitive market good to the sale of a good in a market in which the firm already possessed a dominant share. Since the buyer interested in the good in the monopoly market has limited purchasing alternatives, he is thus forced to purchase the tying good in the competitive market, hence the phrases "extension of monopoly power" and "monopolistic leverage." This theory, however, did not stand up to empirical investigation. Bowman[44] and later Burstein[45] provided evidence refuting this legal interpretation of tie-in sales. Tying arrangements provided the seller with only insignificant market share for the tying good, and thus it was argued that the seller could not be construed as extending his monopoly power into the tied good market.

The goal of tying was not to gain dominant share of the tied good market. The aim, rather, focused on the asset to which this good was tied. Detailed observation of several tying arrangements revealed that the asset was sold at an extremely low explicit price: a price as low as costs could justify. This allowed the firm to attract even the most price-sensitive customers for the asset. The tied good served as a means to an end, it provided a means of monitoring usage intensity for the asset and a way of charging customers on the basis of their use of the asset.

The studies found that the tied good, the means for usage charging, was sold at a premium. The true cost of the asset was therefore the sum of the price premiums the user paid plus the initial low asset price. In this way the seller could charge more for the asset to those who used it more intensely while still capturing the market segment of more price sensitive customers who use it, and value it, less. (Chapter 5 reveals how IBM's practice of tying computer punch cards to mainframes was in fact used to extract consumers' surplus in the mainframe market rather than to extend monopoly power into the card market, as was argued in court.)

Price Segmentation through Bundling

Product bundling, although superficially similar tie-in sales, insofar as the technique involves two products, is fundamentally different in the means by which segmentation is facilitated. Tie-in sales achieve price segmentation in only one product by using the tied product as a device for charging customers on the basis of use intensity for the other product. Product bundling, on the other hand, results in price segmentation in all the markets of the products in the bundle, and no metering device is used.

The key to understanding how bundling can segment customers and result in the charging of customers on the basis of price characteristics[46] lies in what has been labeled the concept of "preference reversal."[47] Stigler first identified the price segmentation rationale for bundling.[48] He demonstrated that bundling can facilitate price segmentation not by simply exploiting absolute differences in price sensitivities for one product but by exploiting relative differences in sets of price sensitivities for two or more products in the bundle. (Chapter 5 shows the applicability of this principle of "preference reversal" to an actual case of bundling in the mainframe industry, thereby revealing the presence of price segmentation.)

While the first identification of bundling as a means to price segmentation is attributed to Stigler, Adams and Yellen developed a comprehensive framework for understanding the price segmentation basis of product bundling in a widely cited essay.[49] They conclude that:

The profitability of commodity bundling can stem from its ability to sort customers into groups with different reservation price characteristics, and hence to extract consumer surplus. . . . In the real world firms cannot always resort to conventional forms of price discrimination in order to extract consumer surplus: reservation prices of specific customers are typically unknown; even if they were known, laws like the Robinson-Patman Act might prevent a seller from using them in an overtly discriminatory scheme. Commodity bundling can overcome these two practical problems associated with conventional price discrimination. In some circumstances, bundling is just as profitable as Pigouvian price discrimination of the first degree. In most circumstances, it is more profitable than simple monopoly pricing.[50]

The essay goes on to distinguish among three bundling policies: (1) unbundled selling, (2) pure bundling, and (3) mixed bundling. Mixed bundling is the practice of offering a bundled package of goods as well as selling them individually. The bundled package is offered at a lower price than the sum of the prices of the goods sold individually. Jeuland points out, "depending on the distribution of reservation prices, any ranking—in terms of profitability for the seller—of these three strategies is possible."[51] In virtually all cases, however, bundling strategies are found to be more profitable than no bundling, and mixed bundling is *generally* more profitable than pure bundling.

In summary, the fundamental contribution of mainstream economics to the idea of customer-dimensional behavior lies in the logic of disaggregated demand. Beyond this, marketing literature suggests economics has contributed in a related way: it set the groundwork for the establishment of explicit criteria for optimal segment definition. A further contribution of economics, however, lies in the analysis of metering, bundling, and tie-in sales as forms of price segmentation.

Limited Links to Product Differentiation

Some literature recognizes limited links between discriminating behavior and differentiating behavior.[52] Dupuit hinted at such links with his early, lost classical terminology of "nonprice discrimination."[53] Pigou saw a link to the extent that product differentiation served as a vehicle for discrimination through labeling, marking, and "all incidents designed to prevent possible purchasers of the grades that are highly priced relatively to the cost of production from becoming, instead, purchasers of the grades that are sold at a lower rate of profit."[54] The less strict economic definition of price discrimination, which has also been adopted by public policy, permits some product variability, but the differences in prices charged for the products must be more than proportional to any cost differences. Here small and/or cosmetic product differences serve to justify higher prices that are disproportionate to any differential in customer benefit. (Chapter 5 shows this is exactly the strategy behind IBM's practice of the "functional pricing" of mainframes.)

Price discrimination analysis, however, was not integrated into the traditional theories of monopolistic competition that treat product differentiation. Chamberlin's work does not consider price discrimination. Robinson's work does consider price discrimination but it is not integrated with her imperfect competition model. It is treated as an isolated analysis upon which "the rest of the book does not depend."[55]

The Applicability of Price Segmentation Analysis to Actual Marketing Behavior

The importance placed on price segmentation as a practical tool of marketing policy is less than that placed on product augmentation, price competition, and especially market segmentation (the central focus of much marketing literature). The role of price segmentation as a marketing practice is discounted because of legal restrictions on discriminatory pricing and the perceived difficulties of implementing segmented pricing that derive from operational and methodological problems in empirically measuring and distinguishing price elasticities among customers.[56] The discounting of the applicability of price segmentation analysis to actual observed behavior betrays a lack of understanding of both the subtle forms segmented pricing may take and the less strict requirements necessary for carrying out these segmented pricing programs relative to implementing Pigouvian-type price discrimination.

While the policy and legislation of many industrial nations inhibits the practice of price segmentation through a qualified prohibition on discriminatory pricing, price segmentation remains a significant form of market behavior.[57] Restrictions on blatant price discrimination do not necessarily affect the more subtle segmented-pricing practices, such as bundling, which are perhaps more prevalent than they would otherwise be were it not for the legal restrictions on more explicit forms of this behavior.

Such forms of segmented pricing price-discriminate only indirectly in so far as they do not entail explicitly offering different prices to different customers. Moreover, the identification of customer groups by determining the differences in price elasticity is not a precondition to implementing these forms of segmented pricings. Tie-in sales, metering, and product bundling exploit customer heterogeneity, and specifically heterogeneity as to price sensitivities, without such measurement. Bundling, for instance, may facilitate the extraction of consumer surplus by a seller without the seller ever knowing the level and differences in price sensitivities among customers.

The conspicuous paucity of references to the recent contributions of business economists in the area of price segmentation by marketing writers suggests a lack of awareness of this work. It is not clear that the applicability of the price segmentation logic of tied sales, metering, and bundling to actual market behavior is fully recognized. As Chang argues:

Despite the limitations of . . . price discrimination models, a valuable opportunity which is often missed is the close parallel between these concepts and the . . . task faced by marketers. In fact, price discrimination theory can act as a bridging mechanism for marketing and microeconomics. This is because price discrimination is one of the few theories in economics devoted to buyer heterogeneities.[58]

Chapter 5 shows the applicability of price segmentation logic to actual behavior in computer mainframe markets.

TWO-DIMENSIONAL MARKET HETEROGENEITY AND TWO-DIMENSIONAL MARKET BEHAVIOR

The Contribution of Marketing Thought from the Perspective of Two-Dimensional Market Heterogeneity

A major contribution of marketing thought in the area of product dimensional behavior was examined in the first section of this chapter in the discussion of product augmentation. The second section briefly referred to marketing work in the area of optimal segmentation, which is directly rooted in the economic logic of price discrimination. The most important contribution of marketing thought to the development of a enlarged framework of market behavior, however, lies in the distinction of product differentiation and the marketing concept of market segmentation and, in particular, the identification of the basis in economic logic for this distinction.

The clear distinction of these two alternative forms of market behavior *implicitly* rests on the distinction of the two dimensions of market heterogeneity: product heterogeneity and customer heterogeneity. The economic basis for this distinction is recognized in the "alternative strategies" model of marketing thought. The contribution of this model to the broader framework called for in this study lies specifically in the identification that "market segmentation" implies an economic logic that is not necessarily implied in product differentiating behavior.

This distinction between product differentiation and market segmentation was recognized by the marketing consultant and scholar Wendell Smith in 1956, who correctly argued that "attempts to distinguish between these approaches may be productive of clarity in theory as well as greater precision in the planning of marketing operations." He reached this conclusion based, in part, on the observation that there were

an increasing number of cases in which business problems have become soluble by doing something about marketing programs and product policies that overgeneralize both markets and marketing effort. These are situations where intensive promotion designed to differentiate the company's products was not accomplishing its objective—cases where failure to recognize the reality of market segments was resulting in loss of market position.[59]

Both the development of the concept of market segmentation and the emergence of its actual use in marketing behavior is explainable in terms of the historical context of the period and, more specifically, the intellectual climate as well as the empirical reality on both the supply side and demand side of the market. The view that the seemingly anomalous market behavior following the Korean War is responsible for the development of market segmentation as a form of marketing behavior and strategy is advocated by Assail:

The concept of market segmentation can be considered the result of the post-Korean War buyer's market. Sometime around 1953–54, it became clear that consumers were holding back on expenditures for durables despite sufficient purchasing power. This was partly due to a stocking up at the start of the Korean War and to a greater sophistication in shopping habits. It became apparent that strategies of convergence of resources on a limited number of offerings could not be sufficient to gain a competitive advantage.[60]

In essence this view argues that the nature of customer purchasing patterns, and specifically the lack of expenditure despite high spending ability, gave rise to market segmentation.

Smith, with no such hindsight, argued for a greater role of market segmentation partly on the basis of changes in sellers' manufacturing technology and, in particular, the ability to achieve scale economies at lower relative levels of production. Smith believed that, although differentiation was, at one time, "essential as the marketing contingent to the standardization and mass production in manufacturing because of the rigidities imposed by production cost considerations," the current state of production methods diminished the need to concentrate all resources on one product to achieve a competitive cost position. The two views (Assail's and Smith's) are not inconsistent, and it seems logical that both demand side and supply side market characteristics contributed to the conceptual development and actual use of market segmentation.

The Logic of Alternative Marketing Strategies
The major contribution of the alternative strategies model lies in the clarification of the differences in the logic of two marketing strategies in the common language of demand analysis. Based on the treatment of discriminatory behavior in Robinson's imperfect competition model, Smith's alternative strategies model posits that, "Segmentation is disaggregative in its effects and tends to bring about recognition of

several demand schedules where only one was recognized before."[61] At a different point the alternative strategies model explains product differentiation in terms of the monopolistic competition model: "In its simplest terms, product differentiation is concerned with the bending of demand to the will of supply. It is an attempt to shift or to change the slope of the demand curve for the market offering of an individual supplier."[62] By juxtaposing these two statements here it becomes clear that Smith's article contained the seeds of a new enlarged framework of strategic marketing behavior.

Apart from this economic logic-based distinction, the work of Smith is also important for making the more general distinction between marketing strategies that appeal to benefits *desired universally* by all customers in a market and marketing strategies that appeal to benefits *desired by only a segment* of customers. The first marketing strategy seeks to make all customers more satisfied; the second marketing strategy seeks to achieve a more precise fit between product offerings and customer needs. This latter, more general, distinction between product differentiation and market segmentation, clarified by Smith, remains an important component of a few business studies scholars. It is recognized by Shapiro who distinguishes between "value-orientated strategies" and "variety-orientated strategies" on the basis of whether they appeal to either "universal benefits" or "diverse specifications" respectively.[63] This latter phraseology originated with Patch[64] (who derived the concepts from Lancaster's work on optimal product variety[65]). Similarly, *implicit* in Porter's three "generic strategies" of "cost," "differentiation," and "focus"[66] is the distinction between universal benefits and segment-specific benefits.

The qualitative difference of the alternative strategies model's contribution, beyond the distinction drawn by the three business scholars above, lies in its roots in the logic of demand analysis. Where these business writers distinguish the strategies simply on the basis of the degree of market focus that a strategy implies (general market versus market segment), the alternative strategies approach roots the distinction in the separate concepts of "turning" or "kinking" the demand curve on the one hand, and "disaggregating" or "tapping" the demand curve on the other hand.

Important Limitations of the Alternative Strategies Model

The major drawbacks of the model lie in (1) its narrow view of what constitutes product differentiation, (2) its incomplete view of the economic logic underlying market segmentation, and (3) the failure to clearly distinguish alterations in the customer's perception of the product and alterations in customer wants themselves. More specifically, concerning the first two points, the alternative strategies framework is built on an unnecessarily narrow conceptualization of product-differentiating behavior and a view of market-segmenting behavior that links it with the economic logic of demand disaggregation only. These two concerns are treated in the following two subsections. The third drawback lies in the fact that Smith does not precisely clarify the difference between changing a customer's perception of a product through either physical changes or advertising, and changing the very nature of customer wants to shape them toward products offered. This distinction is discussed in the next chapter.

A Narrow View of Product Differentiation

Smith views differentiation almost exclusively from the perspective of *promotional strategy* in *consumer goods markets* that has, as its primary tool, *advertising:*

Product differentiation is concerned with the bending of demand to the will of sup-ply . . . variations in the demands of individual consumers are minimized or brought into line by means of effective use of *appealing product claims* designed to make a satisfactory volume of demand converge upon the product or product line being promoted. . . . From a strategy viewpoint, product differentiation is securing a measure of control over the demand for a product by *advertising* or *promoting* differences between a product and the products of competing sellers. . . . Differentiation tends to be characterized by heavy use of *advertising* and *promotion*. . . . It may be classified as a *promotional* strategy.[67] (My emphasis.)

Smith's view can be interpreted as an exaggeration of the biased emphasis in the work of Chamberlin and others toward depicting differentiating behavior with ap-plications to advertising and cosmetic product changes in consumer markets (from which Smith directly adopted his framework). Although Smith's contribution was to link marketing strategy with the Chamberlinian economic concept of product differ-entiation, Chamberlin's view of differentiation was, ironically, strongly influenced by business literature and especially literature on advertising. In Appendix H of the eighth edition of his *Theory of Monopolistic Competition*, in which he discusses "The Origin and Early Development of Monopolistic Competition Theory," Cham-berlin credits his idea, in part, to "substantial reading in the literature of business economics in general, but also with some special reference . . . to the phenomenon of advertising."[68]

The emphasis on product differentiation as an artificial creation of the seller in (at least initially) monopolistic competition theory and exaggerated in the alternative strategies model, may be symptomatic of an earlier noted shortcoming of the prod-uct differentiation framework. Had monopolistic competition theory been linked to price segmentation analysis, the importance of customer heterogeneity (not just product heterogeneity) would perhaps be more fully recognized. The notion of product differentiation as created from or "pushed" from the supplier side would have been balanced by the complementary understanding that product differentia-tion is also "pulled" or demanded from the buyer side because of the heterogeneous nature of customer wants.

In addition to not giving full recognition to the importance of the customer-pull aspect of product differentiation, Smith's view is unnecessarily constrained for another, more obvious reason: it does not acknowledge the nonpromotional sources of product differentiation. The use of the phraseology "bending of demand *to the will of supply* "hints at this major shortcoming. The phrase, "to the will of supply," is suggestive of a supply of goods that is inflexible, toward which demand is bent. Through advertising, consumer goods marketers may be able to bend demand "to the will of supply" without altering supply or in other words without changing actual product characteristics. In this sense the phrase "will of supply" is an accurate metaphor for Smith's version of differentiation. But this version is un-necessarily narrow, for while it is certainly accurate in as far as it recognizes advertising as a means of affecting demand via perception, it does not acknowledge the full diversity of paths to product differentiation outside promotional activity. The utility of the alternative strategies model is limited for this study because it does not acknowledge the importance of augmentation in industrial markets, which for the most part, takes nonpromotional forms.

A Narrow View of the Underlying Logic of Customer Segmentation

The alternative strategies model correctly associates what is termed "market segmentation" with the logic of the disaggregation of demand. However, the alternative strategies model explains segmentation *exclusively* in terms of demand curve disaggregation. The logic of disaggregation does underlie this segmentation, but as the next section explains, it is not the sole logic applicable to this form of market behavior.

This second limitation of the alternative strategies model is, like the first, understandable in light of the existing economic frameworks from which Smith was working. The alternative strategies model, drawing on the price discrimination model, applied its logic without modification or enhancement to the concept of customer segmentation.[69] The alternative strategies model thus sought to explain and distinguish two marketing strategies in terms of two distinct economic frameworks. The result was a framework that conceived marketing behavior (outside pure price competition) in terms of a dichotomy: product differentiation (seen in terms of promotion in consumer markets) versus market segmentation (associated solely with the disaggregation of demand); hence, Smith entitled his article: "Product Differentiation and Market Segmentation As Alternative Marketing Strategies."

Product Segmentation

Smith viewed his marketing strategies as alternatives: a broad based appeal to customers versus one or more focused appeal(s) to customers, with each alternative possessing its distinct and separate economic logic. This book proposes a new view of customer segmentation behavior that implies that segmenting behavior cannot be represented as the alternative to product differentiation logic; rather, it entails the intersection of the logic of exploiting customer heterogeneity and the logic of exploiting potential product heterogeneity. In particular, this book proposes a new category of market behavior: product segmentation. Product segmentation is similar to the modern, generic marketing concept of market segmentation, but it is more precisely defined in terms of economic logic. It is also related to the economic concepts of product differentiation and price discrimination, but it is distinct in that it incorporates, in one concept, the logic elements of both—elements that have previously remained apart in the separate concepts of differentiation and discrimination. Product segmentation is customer segmentation based on product augmentation, or in other words, product augmentation with respect to a customer segment. More briefly put, it is segment-based augmentation and product-based segmentation (hence the term "product segmentation").

The logic of demand curve disaggregation implies only price-segmentation behavior. It does not imply changes along product dimensions. The Chamberlinian logic of twisting the demand curves implies only product augmentation behavior. It does not imply exploiting differences in customers. The concept of product segmentation incorporates elements of both sets of economic logic. It thus has a fuller logic base than the marketing notion of Smith's market segmentation, which explicitly includes only the economic logic of demand disaggregation. (This is illustrated in Figure 2.1.) At the same time, it is a more narrow concept than market segmentation in that it allows only one basis for segmenting customers: the need for a specific product augmentation.

The contemporary notion of market segmentation permits a vast array of segmentation criteria such as the customer's location, the customer's industry, the customer's type of business, the size of the customer's business, the size of the customer's account and so on. As Morierity explains, market segmenting behavior, especially in industrial markets, is traditionally based on easily definable, surface customer differences:

Traditionally industrial markets have been segmented from the seller's viewpoint by grouping buyers on the basis of readily observable characteristics. This approach typically involves segmenting markets on the basis of industry sector, product usage, or such demographic characteristics as company size and geographical location. One very popular approach is to segment an industrial market on the basis of both industry sector and company size.[70]

Unlike such market segmentation, product segmentation admits only one segmentation basis: product augmentation. The common marketing practice of focusing sales efforts on particular locations or particular industries, although a form of *market* segmentation, is not *product* segmentation as developed in this book because it does not segment customers on the basis of product needs. Such segmentation falls short of the potential of product segmentation because it does not fully exploit both dimensions of market heterogeneity. In particular, it does not focus on the product and, more specifically, the product-augmenting logic component of product

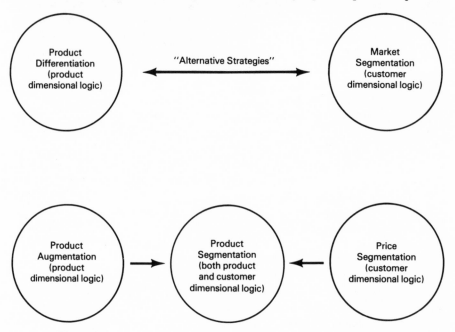

Figure 2.1. Alternative strategies model versus product segmentation.

segmentation. For instance, geographical location may not reveal anything about customer differences as to product augmentations needed.

The notion of "benefit segmentation" recently developed in the marketing literature roughly corresponds to product segmentation developed here. Although the literature on benefit segmentation does not link this concept with the two-dimensional economic logic here suggested and focuses more on statistical tools for identifying "benefit segments," it is similar to the concept of product segmentation insofar as the criterion it advocates to segment customers is essentially the same as that of product segmentation. As Russel Haley, coiner of the term "benefit segmentation," explains:

An approach to market segmentation whereby it is possible to identify market segments by causal factors rather than descriptive factors, might be called "benefit segmentation." The belief underlying this segmentation strategy is that the benefits which people are seeking in consuming a given product are the basic reasons for the existence of true market segments.[71]

While benefit segmentation has been applied to specifically consumer markets as a way of designing an advertising strategy,[72] "there is little evidence that this approach has been used in industrial marketing."[73] This is partly due to the fact that advertising generally plays a much smaller role in industrial markets, as is the case in market for mainframe computers. It is here suggested that the potential contribution the benefit segmentation concept has to offer industrial market strategists (and business strategists in general) lies in its use as the closest existing marketing concept to the idea of product segmentation developed in this study.

Because the demand curve can portray market disaggregation only in terms of heterogeneous price elasticities for otherwise homogeneous customers, it can only represent insulation from customers switching to competitive substitutes in terms of diminished customer responsiveness to price. But product segmentation involves disaggregation on the basis of nonprice attributes: product augmentation needs, not price sensitivities. Product segmentation reduces customers' sensitivity not only to competitive price cutting but also competitors' nonprice product actions. Because this second type of protective insulation is not readily expressed in terms of a demand curve framework, the need for further economic constructs is indicated. Toward this end future research might focus on developing and specifying quantifiable variables of "product elasticity" and "cross-product elasticity." (Chapter 6 shows the idea of insulation from competitors' product-dimensional behavior to be directly applicable to the computer industry. In particular, the chapter explores in detail how the product-segmenting behavior of Cray Research and Tandem Computer has insulated them not only from the price competitive tactics of low-price suppliers such as Fujitsu, but also from the product-augmenting behavior of the dominant general market competitor, IBM.)

The Significance of Economic Logic for the Clarity of Marketing Concepts

The shortcoming in the reasoning of the alternative strategies model revealed in this study is not meant to detract from the model's fundamental insight. The insightful distinction of behavior forms based on the logic of demand is arguably neglected in current marketing literature, with the result that the difference between product differentiation and customer segmentation[74] has become blurred and confused. This point has been very recently confirmed in a comprehensive survey of current marketing literature. Dickson and Ginter report:

Markets have been segmented and products and services differentiated for as long as suppliers have differed in their methods of competing for trade. . . . We therefore might expect that by now the basic purpose, definition of terms, and theory underlying market segmentation and product differentiation would have been consistently described. . . . This is not the case. A review of . . . contemporary marketing textbooks reveals considerable confusion.[75]

The concept of product differentiation, originally implying a turning of the demand curve (universal benefits), has come to mean the much less precise phenomenon of simply offering different products. As the offering of different products is also a manifestation of strategies of customer segmentation (specific benefits), the distinction between product differentiation and customer segmentation does not always remain clear in analysis. A recent doctoral dissertation from the Stanford Business School, for instance, acknowledged Smith's distinction (unusual in itself) but then dismissed it as irrelevant to his subject: "In this article [Smith] introduced a distinction between the strategies of product differentiation and market segmentation . . . we will not make this distinction. Product differentiation will simply be a means of implementing a market segmentation strategy."[76] The author thereby reduced the importance of product heterogeneity to the role of a means to the end of customer segmentation. But clearly, as the alternative strategy models illustrates, product differentiation is important as an end apart from customer segmentation.

Bonoma and Shapiro suggest that "it is perhaps unfortunate that Smith chose [the term] 'product differentiation,' since [it] later developed a meaning substantially different from his apparent intent."[77] Smith, however, did not choose this term; economists did. The different meaning to which these authors refer is more precisely the generalization of the term to mean any product changes (whether having universal appeal or segment-specific appeal) and the removal of the underlying economic logic associated with the term. The result is that the importance of product augmentation as a strategy directed at the general market, as distinguished from a segment-focused strategy, is lost.

Spatial Competition

The lack of clarity in the distinction between product augmentation that appeals to universal benefits and product segmentation that appeals to segment-specific benefits may be symptomatic of the limitation of the *traditional* economic constructs available for explaining such behavior. The traditional demand curve framework treats products as homogeneous commodities and treats customer preferences ("tastes") exogeneously. There is, however, an area of economics outside the mainstream of monopolistic competition theory and conventional price discrimination theory that does allow for the *explicit* modeling of both product heterogeneity and customer heterogeneity in one framework: spatial competition.[78]

The conventional Chamberlinian monopolistic competition model incorporates the impact of product heterogeneity by changing horizontal demand to downward sloping demand. Conventional price discrimination theory considers customer heterogeneity through the disaggregation of demand. These approaches are inadequate, however, where it is required to consider both product and customer heterogeneity simultaneously. For example, product segmentation, that combines both product- and customer-dimensional behavior and specifically segments customers on the basis of product need differences, is not readily modeled in the conventional demand curve framework because demand can be disaggregated only on the basis of price sensitivity. Spatial competition models overcome this limitation and therefore permit the modeling of behavior that simultaneously exploits both dimensions of heterogeneity.

Spatial Competition and Product Differentiation

The origin of "spatial economics" is arguably traceable to Hohan-Heinrich Von Thunen, "the father of location theory."[79] The importance of spatial economics for this study, however, does not lie in its contribution to modern regional science but in its utility as an alternative approach to understanding differentiating behavior. The origin of *this* contribution lies with Harold Hotelling.

Just prior to the publication of Chamberlin's *Monopolistic Competition,* Hotelling published an article that focused on the same fundamental idea as Chamberlin's original work but used a much different approach. Hotelling was concerned with a "feature of actual business behaviour . . . which does not seem to have been taken account of in economic theory."[80] He sought to understand the phenomenon of "groups of buyers who will deal with (a seller) instead of with his competitors despite a difference in price."[81] In other words he was concerned with downward-sloping demand.

To understand the phenomenon of buyer loyalty "despite a difference in price," Hotelling introduced the idea of location for both sellers and customers. The model he developed, in isolation from Chamberlin, was much narrower in scope but more fully worked out in details. Specifically he modeled the behavior of duopolists (selling homogeneous products) making costless locational choices for a single plant in a finite, one-dimensional, geographic market represented by a bounded line: "Main Street." In its literal interpretation, Hotelling's model represents customers and sellers as points along a linear geographic market. The distance between a buyer and a seller represents the transport costs incurred to move the product from the buyer's to the seller's location.

The originality of Hotelling's work however does not lie in the incorporation of space into an economic model but in the use of space to explain downward-sloping demand and thus the effect of product differentiation. Buyers are loyal despite a price difference, not only because customers "live nearer to (the sellers) than others or they have less freight to pay," but also because of "some difference in service or quality." Translating his literal case into an analogy for product differentiation in the apple cider market, Hotelling explains: "the measure of sourness now replaces distance, while instead of transportation costs there are now degrees of disutility resulting from a consumer getting cider more or less different than he wants."[82]

Hotelling hence established the "spatial analogy" whereby the locational case becomes the paradigm for any differentiated market. Geographic space becomes "product space" or "preference space." Geometric points, literally used to represent different seller locations, by analogy represent different product varieties with different combinations of characteristics that the seller actually offers. Points used to represent the location of customers now represent the product variety the customers most prefer or, in terms of two-dimensional space, the peak of the customers' preference contours or customer "ideal points." Distance, which literally represents transport costs, by analogy represents the psychic costs of disutility to a customer resulting from the fact that product varieties actually offered deviate from what the customer most prefers.[83]

The introduction of a *two*-dimensional space to represent product differentiation was first suggested by Chamberlin almost two decades after *Monopolistic Competition* was originally published: "We may speak of economic space instead of literal space and conceive of buyers distributed symbolically throughout a multidimensional area containing all the various aspects of products."[84] Chamberlin did not develop this line of reasoning into a full model but did regard it as a basis for the "reformulation of the theory of monopolistic competition . . . an indication of how I should develop the theory if I were to do it all over again."[85] Such a reformulation represents a significant change, for as Alderson points out: "Chamberlin began with product differentiation and only in his later years came to realize that heterogeneity is inherent on the demand side of the market and is not necessarily created by the supplier."[86]

Chamberlin combined his 1951 spatial approach with his 1933 idea of the "chain linking of markets,"[87] an alternative view of markets to the now standard monopolistic competition theory offered in the bulk of his 1933 work. Unlike standard monopolistic competition that views markets as consisting of differentiated products that are in some sense equally substitutable for one another (as measured by cross-elasticity), in chain markets the price of one variety affects only its more proximate competitors. Chamberlin recognized the two implications of this alternative view of markets in 1933: "A (price) cut by one seller may lead to a smaller reduction by the one next to him and soon dissipate itself without spreading far. Or. . . it might force those nearest to meet it in full, this in turn forcing others and so on indefinitely."[88] Chamberlin's linking of the chain market with the spatial analogy in 1951[89] is the precursor to the "near neighbor" literature, which shows that the ability of a price reduction to spread across the market depends upon the number[90] and proximity[91] of competitors in characteristics space.

Spatial Price Discrimination

A limitation typical of spatial analogy models lies in the restriction of a seller's price behavior to "mill pricing" or "free-on-board" (f.o.b.) pricing whereby the "delivered price" (the price paid by the customer) is the "mill price" (the price at the seller's location) plus the transport costs incurred by the seller in delivery. Under this pricing rule mill prices are assumed to be uniform and the difference in delivered prices is always equal to the difference in transport costs. In the product differentiation analogue of this assumption, the customer pays the mill price for the good actually offered (assumed uniform) plus the loss of utility involved in consuming a product that is not preferred. An alternative analogue proposed by Greenhut, Nopman, and Hung is the customer pays the mill price for the basic commodity plus the cost of customizing the product to his own requirements: the distance corresponding to the degree of customization required.[92]

A fundamental objection to this pricing assumption is that it does not exploit the profit opportunities afforded by the protection of space and thus would not be adopted by profit maximizing firms in this imperfectly competitive market. In particular, differences in the distances between sellers and buyers (or differences in how well products offered match customer preferences) create the potential for "spatial price discrimination." For instance a seller with no close competitors can "overcharge" for transportation to buyers close to him. More technically, as Hobbs explains, "under spatial price discrimination, differences in the delivered prices charged by a firm bear no necessary relationship to differences in transport costs."[93] In terms of product differentiation, for instance, customers now pay variable mill prices plus the usual loss of utility. Alternatively, in terms of the customization example, the producer offers to customize the product, but now the price for customizing does not reflect the actual cost of such customizing.

Theoretically, a seller can discriminate against more distant customers or in favor of more distant customers. But the former case, whereby a "phantom freight" element is charged to more distant customers,[94] gives rise to the potential for resale by more proximate buyers and hence is considered "highly unrealistic"[95] and generally unavailable to the seller. In the latter case the delivered price to distant customers does not fully reflect transport costs; the seller absorbs some element of freight costs. This "freight absorption" is the accepted mechanism for spatial price discrimination and the rate of freight absorption has in fact been used as the measure of the degree of price discrimination.[96] Indeed, since absorption is dependent on freight costs, Leontiff proposed that transportation costs set "the upper limit of interregional discrimination."[97]

A subset of the freight-absorption-type of spatial price discrimination is "uniform delivered pricing" whereby producers absorb all the transportation costs . . . and they post a single uniform (delivered) price for all their customers, regardless of where they are located."[98] In the product space analogue, the seller customizes the product to meet different individual requirements but charges one price.

Horizontal and Vertical Product Differentiation
One of the most important contributions of spatial competition models is the application of the logic of product "characteristics" to an important distinction that provides further rigor and clarity to the distinction between two categories of behavior developed in this thesis. This logic was first articulated in characteristics theory,[99] which is not a spatial competition model in the strict sense in that it uses the geographical space analogy but is nevertheless part of the same broader category of address models of differentiation, which view goods as bundles of characteristics[100] that can be described by coordinates ("addresses") in some continuum.[101] Lancaster is the first to articulate the new logic and terminology for an important and older distinction:

An important distinction needs to be made between vertical and horizontal product differentiation. . . . "The 'product' may be improved, deteriorated, or merely changed". . . . Improvement and deterioration correspond to vertical product differentiation in which the absolute quantities of all characteristics per unit of the good are increased or decreased. [This contrasts with] horizontal . . . product differentiation, in which the various product differentiates vary in specification rather than in quality—the goods are "merely changed."[102]

This logic has been rapidly adopted by, and translated into, spatial analogy models and the literature of address models of product differentiation in general.

In its essential definition, vertical differentiation requires unanimous agreement among customers as to the ordinal ranking of a product so as to indicate if a product is unambiguously "better or worse"[103] than other products. In the terms of product characteristics, the level of all attributes is augmented or lowered. It corresponds to the less rigorously defined marketing phraseology (discussed earlier in "A Narrow View of Product Differentiation") of a "value oriented strategy" that appeals to "universal customer benefits." In the terms of this thesis, vertical differentiation exploits product heterogeneity but not heterogeneity as to customer preferences. Indeed since there is unanimous agreement among customers, preferences for the relevant product benefit are, by definition, homogeneous.

In horizontal differentiation unanimous ranking is not possible: the level of some attributes is increased while the level of others is lowered. Horizontal differentiation corresponds to a "variety-oriented strategy" focusing on "segment-specific benefits." It is distinguished from vertical differentiation by the fact that it *does* exploit heterogeneity as to customer preferences. In an unusually explicit identification of this distinctive feature, its importance in horizontal differentiation is emphasized by Gabszewicz and Thisse:

> Horizontal product differentiation is rooted in *taste differences*. More precisely, the potential customers have *heterogeneous preferences* about the proportion in which the attributes of the product should be combined. A wide range of substitute products can then survive in the same market simply because each of them combines the various attributes of the product in a proportion suitable to a particular *segment of customers*. . . . Each variety has its own circle of customers.[104] [My emphasis.]

This is the same fundamental feature that distinguishes this book's concept of product segmentation from product augmentation. Insofar as *horizontal* differentiation combines an appeal to such customer heterogeneity, with the use of product heterogeneity, it corresponds to this study's concept of product segmentation. Insofar as *vertical* differentiation does not involve the exploitation of differences in customer preferences, but rather exploits product change only, in an appeal to universal benefits, it corresponds to this study's concept of product augmentation.

Limitations

Because spatial competition analysis and characteristics space analysis begin with a fundamental structure that is capable of, and indeed designed to, explicitly model differences in both products and customers, they have much greater potential for capturing the full dimensionality of actual market behavior than conventional non-address approaches. In their current state of development however this potential is not fully realized.

Limitations of both spatial and characteristics models, for the purposes of this book, derive from the restriction of their analyses to horizontal differentiation in the preference domain. Both spatial and characteristics models are limited in that their subject of analysis is by no means a comprehensive representation of the actual behaviors in heterogeneous markets. Specifically, both Hotelling-type and Lancaster-type models are designed for the analysis of horizontal differentiation and not vertical differentiation. Indeed Lancaster used characteristic theory terms to describe vertical differentiation only to clarify that it was not the subject of his analysis: "This book is concerned with the analysis of horizontal rather than vertical product differentiation."[105]

Spatial models have also been revealed to have the capacity to describe vertical differentiation; but vertical differentiation is represented by such an awkward plotting of products and preference points that it only emphasizes that the analysis of such behavior is inappropriate to the design purposes of the model. For example in the simplest spatial model (the one-dimensional Hotelling-type model), vertical differentiation defines the condition where the entire set of customers (preference points) are located on one side of the linear market and the entire set of firms (products offered) are located on the other side, with the complete absence of intermingling. In this extreme condition vertical differentiation exists because there is universal agreement among customers on product ranking by distance (utility). But in fact Hotelling's model is appropriately regarded as the "prototype of horizontal differentiation"[106] because it is exactly this intermingling of preference points and products (and thus differences in product preferences) for which this model is designed and used.

This restricted focus of both spatial and characteristics models is symptomatic of the limited way in which they paramitize customer heterogeneity. Specifically, the explicit modeling of customer heterogeneity is confined to differences in customer preferences. This contrasts with a third category of address models of differentiation, quality differentiation models, that paramatize customer income, an indicator of customer reservation prices.[107] Comparing the quality models against the spatial and characteristics models reveals the limited dimensionality of each type.[108]

The essential difference between the quality versus spatial and characteristics models lies in the different "domains" in which they tend to work: the preference domain versus the income domain. The major stream of address model literature paramatizes preferences. Customer differences are represented in preference differences. Differences in reservation prices are not given an explicit dimension and in fact may be treated as homogeneous by assuming uniform reservation prices as in the Hotelling tradition. Conversely, "there is a strand of the literature in which it is assumed that the variety of consumers stems from their income disparities. . . . A parameter is then introduced to represent an income distribution instead of a variety of tastes, and all consumers have the same utility [preference] function."[109] Indeed preferences are treated as homogeneous in so far as the models require universal agreement among customers as to products preferred. Moreover these two approaches are not parallel models simply working in different domains. The first stream that paramatizes preferences does not require that product be rankable by customer's preferences. The second, smaller strand that paramatizes income not only requires that products be rankable by universally agreed upon preferences (vertical differentiation) but also that customers be rankable by their reservation prices (the "monotonicity assumption"). Thus this second strand is inconsistent with the basic preference-paramatizing model (Hotelling's model) in that monotonicity contradicts the model's assumption as to reservation prices.

Ideally the full dimensionality of market behavior could be best captured by a model that simultaneously paramatizes both reservation prices and preferences. Moreover, ideally such a model would be able to simultaneously represent not only pure horizontal and pure vertical differentiation but also differentiation behavior where the logic of both pure cases interact. A recent comment by Shaked and Sutton shows that this limitation has been recognized in the literature.

A striking feature of the existing literature is the fact that the results obtained in the cases of the "pure vertical" and "pure horizontal" differentiation literatures are sharply different, but the empirically relevant case in which both types of attributes are present, has not been widely studied. . . . A general theory of product differentiation, then, requires an analysis of situations in which products may be differentiated both "horizontally" and "vertically."[110]

Resolution of these sharply differing results and the development of a general model that addresses the empirically relevant case is an important subject for further theoretical research. More broadly, in their current state of development, the address models of heterogeneous markets do not comprise a unified theory. The assumptions, mechanisms, results, and conclusions (when they are reached) can vary between preference and income models, between different categories of preference models, as well as within each of these categories of address models.

The Need for a General Framework

Address models of differentiation, like traditional dèmand analysis and even more than traditional economic analysis, have significant potential for bridging economic theory and marketing strategy. Indeed, as Capozza and Van Order explain, the marketing "researchers have developed both the theory and the necessary algorithms for implementation [of market segmentation programs rooted in an approach] closely related to the theory of consumer demand developed by Lancaster."[111] Ironically as Capozza and Van Order further point out, this development has occurred "quite independent of the literature of economics." The lack of attention to the potential for integration of these two specific subfields is neither a new nor unique disappointment but in fact reflects an older and more general tendency of the drifting apart of economics and marketing. Wendell Smith was writing in a period when the contribution of Chamberlin and Robinson only recently brought economic theory closer to market reality. The intellectual climate was ripe for bridging the disciplines of economics and marketing. But the seeds of a general framework of marketing behavior based on economic logic, originally sown by Smith, never took root.

Contributions in macroeconomics and the increasing emphasis on building models requiring ever-limiting assumptions deflected attention from further potential contributions toward linking economic models of behavior with actual market behavior and linking economic logic with marketing strategy.[112] Lancaster criticizes the economic perspective on consumer behavior for isolation from market reality through narrowing assumptions:

The theory of consumer behaviour . . . is a thing of great aesthetic beauty, a jewel set in a glass case. The product of a long process of refinement from the nineteenth century utility theorists through Slutsky and Hicks-Allen to the economist of the last 25 years, it has been shorn of all irrelevant postulates so that it now stands as an example of how to extract the minimum of results from the minimum of assumptions. To the process of slicing away with Occam's razor, the author made a small contribution.[113]

At the same time marketing's increasing emphasis on developing taxonomies of observed marketing phenomena versus the development of conceptual frameworks and generalizations of observed behavior[114] kept the focus of its attention away from the highly abstract models of economics. Lambert, a marketing scholar, suggests that economic thinking is so far abstracted from reality that it is irrelevant to understanding actual behavior and "out of place in the marketing literature."[115] Although recently economic thinking has been fruitfully applied to several specific marketing problems,[116] this work has not been orientated toward providing a common analytic base for the development of a general framework. To the neglect of developing a general framework Smith himself contributed. As Winter argues:

Professor Smith provided a conceptual base in 1956 that is still relevant today. It was unfortunate that he assumed administrative duties and did not remain active in the area. Regarding his contribution, I might quote what a tennis coach once said about my tennis stroke, a great start but no followthrough.[117]

About a decade after Smith's contribution, a very few economics-orientated marketing scholars recognized this drifting apart of the two fields and tried to convey its import for inhibiting the development of a general framework to the bulk of members in their intellectual community. Grether, one of the most outspoken writers on the need to rectify this drift, summarizes the state of affairs:

A large portion, perhaps the bulk, of marketing literature is explicitly written in terms of a narrowly applied interest. Thus [in] Bartels . . . the relatively practical and narrowly descriptive emphasis dominates. . . . Typical marketing textbooks increasingly became encyclopedic collections of statistical data and of descriptive interpretation. . . . Gradually, the use of cases, or of shorter problems, became part of the teaching procedures in most institutions, . . . de-emphasizing formal efforts at "theory" and "generalization."
Economic analysis typically was not and could not be integrated within the taxonomic, eclectic framework of marketing literature. Consequently, economists seeking illumination from marketing literature for the most part found little or nothing that would relate to the accepted categories and theories of competition. [Conversely] the descriptive, taxonomic, and eclectic managerial approaches to some extent led [marketing] away from economic analysis. This was especially so because the taxonomies of the descriptive marketing literature were the natural ones of the real world and not the behavioural ones of the analysis of competition in economics. . . . A consequence of this state of affairs was that marketing teachers and scholars increasingly became cut adrift from economic theory, and hence became disinterested in its use, refinement, or creative adaptation and development. . . . The literature and results both in economics and in marketing might have been vastly different if this type of differentiation had not occurred.[118]

Consistent with such interpretations of intellectual history, predictions for the future of marketing are typically pessimistic about the possibility for general theory. Halbert thus reports:

Most theorists, however, seem to hold less hope for the development of a useful general theory of marketing. They seem to agree that, because of the diversity and complexity of the variables, and the dynamic nature of marketing phenomena, a number of theories about specific aspects of marketing will probably be developed rather than a general theory of marketing.[119]

Among marketing scholars who have expressed a need for a more integrated general theory, a few have also emphasized the need to bridge marketing thought with economics. In the English-speaking world, notable advocates of the development of a general theory of market behavior rooted in economics have been Alderson and Grether.[120] The potential need for a general theory of marketing behavior has also been recognized in the Scandinavian marketing literature in which a nexus with economics has been markedly stronger than in the English language marketing literature. Holback-Hanssen, a marketing scholar as well as practitioner, argued that there is "a need for a more constant and complete theory of marketing" and that there is "a gap between the economic theories of the firm, the current literature on marketing practices and actual behaviour."[121] Similarly Martin Shubik argued that there is a "gap between economic theory, institutional studies and the applied study of the functioning of markets."[122]

What interest that had been generated by marketing scholars such as Smith and Alderson in developing a unified framework of marketing behavior very rapidly diminished as marketing thought became increasingly dispersed into separate focused areas. Hunt and Hunt explain: "Unfortunately, after a spurt of interest in the subject in the 1950s and early 1960s, there was a period of time when very few marketing academicians were actively involved in attempting to theorize concerning marketing."[123] But recently this has changed. The abundance of articles in current marketing periodicals betrays the emerging, widespread recognition among contemporary scholars of the need for an underlying unifying framework and the development of a general marketing theory. As Hunt and Hunt point out:

Interest in the area of marketing theory has grown tremendously in the last few years. There have been several special conferences on marketing theory sponsored by the American Marketing Association, a special plea by the editor of the Journal of Marketing for more theoretically oriented articles, and the American Marketing Association is planning another special conference on marketing theory.[124]

Recognition of the need for a general framework is scattered throughout some of the most recent literature. For instance, Sirgy calls for a new "conceptualization of the consumer behaviour discipline."[125] Arndt points to the need for a "liberating paradigm" in marketing science[126] and suggests "the political economy paradigm" as a "foundation for theory building in marketing."[127] Matsusoki shows "the need for theory development at the macro-marketing level." He argues "the need to develop an adequate theory at the macro-marketing level has increased [even though] no such theory seems to exist."[128] Zaltman reviews the potential for "theory construction" in the marketing discipline.[129] Gaski calls for a general theory of marketing based on "information-power structure" analysis and specifies the requirements needed to build scientific marketing theory.[130] Bartels summarizes the concern faced by scholars of an eclectic discipline characterized by very separate, highly focused, analytic frameworks: "the problem is that our field of study has been too narrow to provide the generalizations required for meeting global needs."[131]

CHAPTER SUMMARY

This chapter drew on the demand logic of economics and the perspective of marketing to develop generic business strategy concepts of market behavior, which later chapters show to be useful in understanding actual behavior. The first section developed the business strategy concept of "product augmentation," which meshes the distilled economic logic of product differentiation with the customer-benefits perspective of the augmented-product concept in the marketing literature. The second section developed the idea of "price segmentation," which draws on the classical economic logic of price discrimination but which is shown to have applicability beyond explicit discriminatory *price* policies to the more pervasive forms of marketing *product* policies such as bundling. The third section presented the concept of "product segmentation," which draws on the economic logic of both product differentiation and price discrimination, thereby providing a fuller and yet more well defined logic base than that associated with the less precise marketing concept of market segmentation.

Using the concept of two-dimensional market heterogeneity, the next chapter brings together these three categories within one framework rooted in the common language of demand logic, supplemented by the customer effectiveness perspective of marketing. More specifically, the next chapter outlines the essential features of a new framework of market behavior and its significance for describing the function of markets and for understanding market strategy. The next chapter does not synthesize a general theory of marketing nor a new economic model of market conduct; rather it presents an intellectual skeleton, which provides insights into the fundamental anatomy of market behavior and from which the substance of an interdisciplinary theory can ultimately be fleshed out. The three chapters following it show the applicability of this market behavior anatomy to actual, observed activity in the mainframe computer industry.

NOTES

1. E. T. Grether, "Chamberlin's Theory of Monopolistic Competition and the Literature of Marketing," in *Monopolistic Competition Theory: Studies in Impact. Essays in Honour of Edward H. Chamberlin,* edited by R. E. Kuene (New York, Wiley, 1967), p. 308.

2. T. Nagle, "Economic Foundations for Pricing," *The Journal of Business,* 57, No. 1, Pt. 2 (January, 1984), pp. 3–4.

3. R. E. Caves, P. J. Williamson, "What is Product Differenitation, *Really?*" *Journal of Industrial Economics,* 34, No. 2 (December, 1985), p. 113.

4. P. A. Samuelson, "The Monopolistic Competition Revolution", in *Monopolistic Competition Theory,* edited by R. Kuenne (New York, Wiley, 1967), pp. 103–138.

5. T. V. Atwater, "'Lost' or Neglected Components of a General Equilibrium Theory of Marketing,"

in *Conceptual and Theoretical Developments in Marketing,* edited by P. C. Ferrell, S. W. Brown, C. W. Lamb, Jr. (Chicago, American Marketing Association, 1979), p. 186.

6. R. D. Peterson, "Product Differentiation, Implicit Theorizing and the Methodology of Industrial Organization," *Nebraska Journal of Economics and Business,* 19 (Spring, 1980), p. 22.

7. Although the terms monopolistic competition and imperfect competition are often used interchangeably, it is monopolistic competition that is more precisely relevant to product heterogeneity. As Chamberlin explains, pure competition is a "much simpler and less inclusive concept than 'perfect' competition." Robinsonian "perfect competition" may imply an "absence of friction in the sense of an ideal fluidity or mobility of factors" such that "adjustments to changing conditions . . . are accomplished instantaneously." Perfect competition may also involve "perfect knowledge . . . and the consequent absence of uncertainty." Pure competition merely presumes many sellers and homogeneous products. Chamberlin thus explains that "'pure' and 'perfect' competition must not be identified; and to consider the theory of monopolistic competition vaguely as a theory of 'imperfect' competition is to confuse the issues." See E. H. Chamberlin, *The Theory of Monopolistic Competition: A Reorientation of the Theory of Value,* 7th ed. (Cambridge, MA, Harvard University Press, 1965), pp. 6. 7.

8. Ibid., p. 7.

9. F. M. Scherer, *Industrial Market Structure and Economic Performance* (Chicago, Rand McNally, 1980), p. 385.

10. P. R. Dickson, J. L. Ginter, "Market Segmentation, Product Differentiation, and Marketing Strategy," *Journal of Marketing,* 51, No. 2 (April, 1987), p. 4.

11. M. J. Baker, *Marketing New Industrial Products* (London, MacMillan, 1976), p. 12.

12. Ibid.

13. H. Demsetz, *Economic, Legal and Political Dimensions of Competition: Professor Dr. F. DeVries Lectures in Economic Theory* (New York, North-Holland, 1982).

14. Samuelson, "The Monopolistic Competition Revolution," p. 113.

15. Ibid.

16. Chamberlin, *Monopolistic Competition,* p. 63. Chamberlin was not the first to suggest this casting, but he was the first (along with Robinson) to develop a model based on this cast. Prior to Chamberlin's seminal work, Knight refers to the "fact that every business is a partial monopoly." He further writes "it is remarkable that the theoretical treatment of economics has related so exclusively to complete monopoly and perfect competition," thus setting the stage for the development of intermediate modes of behavior along this linear continuum: F. H. Knight, *Risk, Uncertainty and Profit* (Boston, Houghton Mifflin, 1921), p. 193. The notion of differentiation as a monopoly element is earlier found in Veblen's identification of differentiation as a key ingredient of industrial success. He argues that "it is very doubtful if there are any successful business ventures . . . from which the monopoly element is wholly absent" (Veblen, *The Theory of Business Enterprise,* p. 54).

17. See W. G. Shepard, *Market Power and Economic Welfare* (New York, Random House, 1970), pp. 48–66, 195–202.

18. W. Alderson, *Marketing Behavior and Executive Action: A Functional Approach to Marketing Theory* (Homewood, IL, Irwin, 1957), p. 103.

19. Atwater, "'Lost' or Neglected Components of a General Equilibrium Theory of Marketing," p. 185.

20. Alderson, *Marketing Behavior and Executive Action,* p. 103.

21. Chamberlin never retracted his position that the price competition model represented an "ideal" point in the *theoretical* sense and that product differentiating behavior existed wholly within the unidimensional continuum between price competition and monopoly as an unpure mixing of both endpoints (rather than existing beyond these traditional frameworks as a pure, new form of behavior no less competitive than price competition). By the late 1940s Chamberlin did, however, recognize that the implications of differentiating behavior challenged the assumptions of welfare economics and reached the conclusion that pure price competition as an "ideal" type in the *welfare* sense was no longer tenable: "the explicit recognition of product differentiation brings into the open the problem of variety and makes it clear that *pure competition may no longer be regarded as in any sense an 'ideal' for purposes of welfare economics*" (emphasis as original). (Chamberlin, *The Theory of Monopolistic Competition,* 6th edition, 1948, pp. 214–215.) But as he points out in the eighth edition, "the welfare problem was not part of my original objective at all" (p. 296). A fuller statement of this emerging welfare position is found in Chamberlin, "Product Variety and Public Policy," *American Economic Review* 40 (May, 1950),

pp. 85–92, which was further developed by L. Abbott, *Quality and Competition: An Essay in Economic Theory* (New York, Columbia University Press, 1955), a precursor to the limited modern economic literature on optimal product variety.

22. See for instance B. P. Shapiro, *Industrial Product Policy: Managing the Existing Product Line* (Cambridge, MA, Marketing Science Institute, 1979); T. V. Bonoma, B. P. Shapiro, *Segmenting the Industrial Market* (Lexington, MA, Lexington Books, 1984); R. T. Morierity, *Industrial Buying Behavior* (Lexington, MA, Lexington Books, 1983); E. F. Webster, Jr., *Industrial Marketing Strategy* (New York, Ronald Press, Wiley, 1979).

23. T. Levitt, *Marketing for Business Growth*, 2d ed. (New York, McGraw Hill, 1974), pp. 34–51.

24. T. Levitt, *The Marketing Imagination* (New York, Free Press, 1983), pp. 72–93; alternatively, see "Marketing Success Through Differentiation—of Anything," *Harvard Business Review* (January–February, 1980), by the same author.

25. R. E. Corey, *Industrial Marketing* (Englewood Cliffs, NJ, Prentice-Hall, 1983), pp. 40–41.

26. R. N. Cardoza, *Product Policy* (New York, Addison-Wesley, 1985), p. 1.

27. J. L. Forbus, N. T. Mehta, "Economic Value to the Customer" (New York, McKinsey and Company Staff Paper, February, 1979).

28. Bonoma, Shapiro, *Segmenting the Industrial Market*, p. 78.

29. M. E. Porter, *Competitive Advantage: Creating and Sustaining Superior Performance* (New York, Free Press, 1985), p. 130.

30. T. Levitt, *The Marketing Mode: Pathways to Corporate Growth* (New York, McGraw Hill, 1982), p. 1.

31. Levitt, *The Marketing Imagination*, p. 77.

32. Grether, "Chamberlin's Theory of Monopolistic Competition and the Literature of Marketing," pp. 307–328.

33. See for instance the definitive history of marketing thought, R. Bartels, *The History of Marketing Thought* (Columbus, OH, Grid, 1976).

34. Grether, "Chamberlin's Theory of Monopolistic Competition and the Literature of Marketing," p. 308.

35. In fact, price segmentation analysis is traceable to one of the co-developers (along with Cournot) of marginal utility theory, Jules Dupruit, whom Edgeworth regarded as "the highest authority on the theory of discrimination" (F. Y. Edgeworth, *Mathematical Physics*, 1881). Economic literature, however, generally associates Marshall's student, Pigou, with the major role in the early development of price segmentation analysis.

36. J. B. Long, Jr., "Comments" on "Gaussian Demand and Commodity Bundling," *Journal of Business* (January, 1984), 57, No. 1, Pt. 2, p. 235.

37. Scherer, *Industrial Market Structure*, p. 17.

38. A. Marshall, *Principles of Economics* (London, Macmillan, 1920), p. 124.

39. A. C. Pigou, *The Economics of Welfare*, 4th edition (London, Macmillan, 1960).

40. J. Robinson, *The Economics of Imperfect Competition* (London, MacMillan, 1948), p. 186.

41. H. Assail, "Demand Criteria for Normative Market Segmentation Theory: A Retrospective View," in *Proceedings of the 11th Paul D. Converse Symposium*, edited by D. M. Gardner, F. W. Winer (Chicago, American Marketing Association, 1982), p. 10.

42. W. F. Sharpe, *The Economics of Computers* (New York: Columbia University Press, 1969), p. 252.

43. *Times-Picayune Publishing v. United States*, 345 U.S. 594, 611 (1953).

44. W. S. Bowman, "Tying Arrangements and the Leverage Problem," *Yale Law Journal*, 67 (November 1957), pp. 19–36.

45. M. L. Burstein, "The Economics of Tie-in Sales," *Review of Economics and Statistics*, 27 (February, 1960), pp. 68–73. See also "A Theory of Full Line Forcing," *Northwestern University Law Review*, 55 (March–April, 1960), pp. 62–95.

46. This section focuses on the recent contribution of economic analysis in identifying a price segmentation logic to the generic practice of product bundling. Chapter 5 reveals that this logic is only one of several ways in which IBM's bundling behavior extracts consumer surplus.

47. R. Schmalensee, "Gaussian Demand and Commodity Bundling," *The Journal of Business*, 57, No. 1, Pt. 2 (January, 1984), p. 211.

48. G. J. Stigler, "*United States v. Lowe's Inc:* A Note on Block Booking," *Supreme Court Review*

(1963), pp. 132–157. Reprinted in G. J. Stigler, *The Organization of Industry* (Homewood, IL, Irwin, 1968). While Stigler is credited with the first clear recognition of bundling as price segmentation, elements of this logic appear in M. L. Burstein, "The Economics of Tie-In Sales," *Review of Economics and Statistics* (February, 1960), pp. 65–73.

49. W. J. Adams, J. L. Yellen, "Commodity Bundling and the Burden of Monopoly," *Quarterly Journal of Economics* (August, 1976), pp. 475–498.

50. Ibid., p. 475.

51. A. Jeuland, "Comments on Gaussian Demand and Commodity Bundling," *The Journal of Business* (January, 1984), 57, Pt. 2, p. 233.

52. Among the most important original contributions to linking price discrimination with superficial forms of product differentiation such as cosmetic differences that prey on consumer ignorance is D. S. Watson, *Price Theory and its Uses* (Boston, Houghton Mifflin, 1968), p. 325, who identified the phenomenon of branded versus generic products as a basis for discrimination. Prior to this, Machlup specified a category of price discrimination termed "product discrimination" in which the seller "appeals-to-the-classes" by charging different prices that are more than proportional to differences in costs: F. Machlup, "Characteristics and Types of Price Discrimination," found in Universities—National Bureau For Economic Research, *Business Concentration and Price Policy* (Princeton, Princeton University Press, 1955), pp. 400–423. A notable portion of Machlup discrimination typology was derived from an even earlier work, R. Cassidy, Jr., "Techniques and Purposes of Price Discrimination," *Journal of Marketing* (October, 1946), pp. 135–140.

53. J. Dupuit, *On the Measurement of Utility of Public Works* (Paris, Carilian-Goeury et V. Dalmont, 1842).

54. Pigou, *The Economics of Welfare*, p. 277.

55. Robinson, *The Economics of Imperfect Competition*, p. 179.

56. Assail, "Demand Criteria for Normative Market Segmentation Theory," pp. 12–16.

57. See for instance M. L. Greenhut, "Spatial Pricing in the United States, West Germany and Japan," *Economica* 48, No. 1, pp. 79–86.

58. D. R. Chang, "Models of Market Separation and Control of Intra-Firm Leakage Effects" (unpublished paper, Harvard Business School, 1985), p. 37.

59. W. R. Smith, "Product Differentiation and Market Segmentation as Alternative Marketing Strategies," *Journal of Marketing* (21 July 1956), pp. 4–5. Also see a restatement of this article with no significant alterations, W. R. Smith, "Product Differentiation and Market Segmentation: Another Look," in *Converse Symposium* (1982), p. 4.

60. Assail, "Demand Criteria for Normative Market Segmentation Theory," p. 9.

61. Smith, "Product Differentiation and Market Segmentation as Alternative Marketing Strategies," p. 5.

62. Ibid.

63. B. P. Shapiro, "Industrial Product Policy Research Project" (unpublished notes, Harvard Business School, 3 April, 1984), pp. 1, 3, Exhibit 1.

64. P. Patch, "Buyer Sophistication, Organization Context and Vendor Selection in the Small Computer Market" (unpublished manuscript, Harvard University, October, 1983).

65. K. Lancaster, *Variety, Equity and Efficiency: Product Variety in an Industrial Society* (New York, Columbia University Press, 1979).

66. M. E. Porter, *Competitive Strategy: Techniques for Analyzing Industries and Competitors* (New York: Free Press, 1980), pp. 35–40.

67. Smith, "Product Differentiation and Market Segmentation as Alternative Marketing Strategies," pp. 4–6.

68. Chamberlin, *Monopolistic Competition*, pp. 292–318.

69. Actually Smith used the two chapters on price discrimination from Robinson's work and thus technically based his work on only the imperfect and monopolistic competition literature, although as noted previously price discrimination logic clearly derives from earlier sources.

70. R. Morierity, *Industrial Buying Behavior* (Lexington, MA, Lexington Books, 1983), p. 93.

71. R. Haley, "Benefit Segmentation: A Decisions Oriented Tool," *Journal of Marketing*, 32 (1968), pp. 30–35.

72. R. Haley, "Beyond Benefit Segmentation," *Journal of Advertising Research* (1971), pp. 3–8.

73. Morierity, *Industrial Buying Behavior*, p. 94.

74. The term "customer segmentation" is used here as a broad category that includes both the traditional idea of market segmentation and the concept of product segmentation developed in this book.

75. Dickson, Ginter, "Market Segmentation, Product Differentiation, and Marketing Strategy," p. 4.

76. K. S. Moorthy, "Market Segmentation Through Product Differentiation" (Stanford Business School Ph.D. thesis, 1983), p. 198, available from University Microfilms, Ann Arbor, MI.

77. Bonoma, Shapiro, *Segmenting the Industrial Market*, p. 3.

78. The characterization of spatial economics as outside the "mainstream" is common among several historians of economic thought as well as current theorists in the area. For instance, historian Blaug describes "the continued neglect of spatial economics by mainstream economists [that] largely continues to this day." Theorist Dorward explains that "the development of spatial theory has been largely isolated from mainstream economics and confined to a number of areas of special interest." Hence, Norman concludes spatial economists have had only "a marginal impact on mainstream economics." Prominent theorists Gabszewicz and Thisse lament that "the field of spatial competition is more central to economic theory than its peripheral status would suggest." As to spatial price-discrimination theory in particular, prominent theorists Greenhut and Ohta similarly argue: "Spatial price discrimination should have become an integral part of the subject at a relatively early date in the development of price discrimination theory, such was not the case. The analysis of spatial price discrimination followed that of nonspatial price discrimination by about 50 years; and even today it is largely discussed by only a rather small group of economists." References for quotations in the order cited are as follows: M. Blaug, *Economic Theory in Retrospect* (Cambridge, Cambridge University Press, 1985), p. 629; N. Dorward, "Recent Developments in the Analysis of Spatial Competition and Their Implications for Industrial Economics," *Journal of Industrial Economics*, 31, Nos. 1–2 (September–December, 1982), p. 133; G. Norman, *Spatial Pricing and Differentiated Markets, London Papers in Regional Science*, 16 (London, Pion, 1986), p. 2; J. J. Gabszewicz, J. F. Thisse, "Spatial Competition and the Location of Firms" in *Location Theory*, edited by J. J. Gabszewicz, J. G. Thisse, M. Fujita, U. J. Schweizer (Chur, Switzerland, Harwood Academic Publishers, 1986), p. 61; M. L. Greenhut, H. Ohta, *Theory of Spatial Pricing and Market Areas* (Durham, NC, Duke University Press, 1975), p. 39.

79. C. Ponsard, *History of Spatial Economic Theory* (Berlin, Springer-Verlag, 1983), p. 10. This is a revised, translated edition of the original 1958 French edition.

80. H. Hotelling, "Stability in Competition," *Economic Journal*, 39 (March, 1929), p. 44.

81. Ibid.

82. Ibid.

83. See Appendix B for later developments of this approach.

84. E. H. Chamberlin, "Monopolistic Competition Revisited," *Economica* (November, 1951), p. 347.

85. Ibid., p. 343.

86. Alderson, *Marketing Behavior and Executive Action*, p. 103.

87. Chamberlin, *Monopolistic Competition* (1933), p. 103.

88. Ibid., p. 104.

89. Chamberlin, "Monopolistic Competition Revisited," p. 355: "the typical chain relationship . . . emerges so clearly from our earlier spatial example."

90. G. C. Archibald, G. Rosenbluth, "The 'New' Theory of Consumer Demand and Monopolistic Competition," *Quarterly Journal of Economics*, 89 (1975), pp. 569–90.

91. R. Rothschild, "Competitive Behaviour in Chain-Linked Markets," *Journal of Industrial Economics*, 31, Nos. 1–2 (September–December, 1982), pp. 57–67.

92. M. L. Greenhut, G. Nopman, C. Hung, *The Economics of Imperfect Competition: A Spatial Approach* (Cambridge, Cambridge University Press, 1987), p. 293.

93. B. F. Hobbs, "Mill Pricing Versus Spatial Price Discrimination Under Bertrand and Cournot Spatial Competition," *Journal of Industrial Economics*, 35, No. 2 (December, 1986), p. 175.

94. D. Dewey, "A Reappraisal of F.O.B. Pricing and Freight Absorption," *Southern Economic Journal*, 22 (July, 1955), pp. 48–54.

95. Greenhut, Ohta, *Theory of Spatial Pricing*, p. 43.

96. This idea was first formulated by E. M. Hoover in "Spatial Price Discrimination," *Review of Economic Studies*, 63 (1949), pp. 289–314.

97. W. Leontief, "The Theory of Limited and Unlimited Discrimination," *Quarterly Journal of Economics,* 54 (1939), p. 491.

98. R. E. Schuler, B. F. Hobbs, "Spatial Price Duopoly Under Uniform Delivered Pricing," *Journal of Industrial Economics,* 31, Nos. 1–2 (September–December, 1982), p. 175.

99. K. J. Lancaster, "A New Approach to Consumer Theory," *Journal of Political Economy,* 74 (1966), p. 133.

100. D. A. Hay, D. J. Morris, *Industrial Economics Theory and Evidence* (Oxford, Oxford University Press, 1979), p. 84.

101. D. Neven, "Address Models of Differentiation," *London Papers in Regional Science,* 16 (1986), p. 5.

102. K. J. Lancaster, *Variety, Equity and Efficiency: Product Variety in an Industrial Society* (New York, Columbia University Press, 1979), pp. 27, 28.

103. In this way vertical differentiation corresponds to what has earlier been called "quality change" M. Mussa, S. Rosen, "Monopoly and Product Quality," *Journal of Economic Theory,* 18 (1978), pp. 301–317. However, "quality" is sometimes used in a sense that does not necessarily imply vertical ranking: N. E. Leland, "Quality Choice and Competition," *American Economic Review,* 67, pp. 127–135.

104. J. J. Gabszewicz, J. F. Thisse, "On the Nature of Competition with Differentiated Products," *Economic Journal,* 96 (March, 1986), p. 160.

105. Lancaster, *Variety, Equity and Efficiency,* p. 28.

106. Gabszewicz, Thisse, "On the Nature of Competition with Differentiated Products," p. 161.

107. Note also that there are significant inconsistencies between the two categories of address models that paramatize preferences. See, for instance, Archibald, Eaton, and Lipsey who conclude that "it is false to assume that any characteristics model necessarily 'twins' with all, or indeed any, spatial model, or vice versa." G. C. Archibald, B. C. Eaton, R. G. Lipsey, "Address Models of Value Theory," *New Developments in the Analysis of Market Structure, Proceedings of a Conference Held by the International Economic Association,* edited by J. E. Stiglitz, G. F. Mathewson (Cambridge, MA, MIT Press, 1986), p. 38.

108. There is another problem of restricted focus that is common to address models in general. In particular, consistent with Chamberlinean monopolistic competition and much marketing research, the-oretical and applied address-model research focuses on consumer goods. For example, in one of the most comprehensive articles discussing all three categories of address models readers are cautioned in the very first paragraph that "we confine ourselves throughout to the market for consumer goods (Chamberlin's 1933 problem) and do not consider . . . intermediate goods." Archibald, Eaton, Lipsey, "Address Models of Value Theory," p. 3.

109. Neven, "Address Models of Differentiation," p. 7.

110. A. Shaked, J. Sutton, "Product Differentiation and Industrial Structure," *Journal of Industrial Economics,* 36, No. 2 (December, 1987), pp. 132, 134.

111. D. R. Capozza, R. Van Order, "Product Differentiation and the Consistency of Monopolistic Competition": A Spatial Perspective," *Journal of Industrial Economics,* 31, Nos. 1–2 (September–December, 1982), p. 30.

112. Atwater, " 'Lost' or Neglected Components of a General Equilbibrium Theory of Marketing," p. 184.

113. Lancaster, "A New Approach to Consumer Theory," p. 130.

114. Grether, "Chamberlin's Theory of Monopolistic Competition and the Literature of Marketing," p. 308.

115. D. R. Lambert, "On Paying Homage to a False God: Comments on the Theory of the Firm's Role," in *Conceptual and Theoretical Developments in Marketing,* edited by O. C. Ferrell, S. Brown, C. Lamb (Chicago, American Marketing Association, 1979), pp. 363–373.

116. Indeed the use of economic analysis to illuminate and solve specific marketing issues is a theme of many recent articles in *The Journal of Marketing Research* and *Marketing Science;* for instance, see R. J. Dolan's treatment of quantity discounting: "Quantity Discounts: Managerial Issues and Research Opportunities," *Marketing Science* (Winter, 1987).

117. F. W. Winter, "Market Segmentation: A Review of its Problems and Promise," in *Proceedings*

of the 11th Paul D. Converse Symposium, edited by D. M. Gardner, F. W. Winter (Chicago, American Marketing Association, 1982), p. 28.

118. Grether, "Chamberlin's Theory of Monopolistic Competition and the Literature of Marketing," pp. 308, 314, 315.

119. M. Halbert, *The Meaning and Sources of Marketing Theory* (New York, McGraw Hill, 1965), p. 66.

120. Grether's most important work in this area is R. S. Vaile, E. T. Grether, R. Cox, *Marketing in the American Economy* (New York, Roland Press, 1952).

121. L. Holback-Hanssen, *Contributions to a Theory in Marketing* (New York, Wiley, 1958), p. 1.

122. M. Shubik, *Strategy and Market Structure* (New York, Wiley, 1959), p. xii.

123. S. D. Hunt, K. A. Hunt, "Bartel's Metatheory of Marketing: A Perspective," in Winter, Gardner, eds., *Converse Symposium,* p. 50.

124. Ibid.

125. M. Sirgy, "A Conceptualization of the Consumer Behavior Discipline," *Journal of the Academy of Marketing Science,* 13 (Winter–Spring, 1985), pp. 104–121.

126. J. Arndt, "On Making Marketing Science More Scientific: Role of Orientations, Paradigms, Metaphors and Puzzle Solving," *Journal of Marketing,* 49 (Summer, 1985), pp. 11–23.

127. J. Arndt, "The Political Economy Paradigm: Foundation for Theory Building in Marketing," *Journal of Marketing,* 47 (Fall, 1983), pp. 44–45.

128. H. Matsusaki, "Marketing, Culture and Social Framework: The Need for Theory Development at the Macro Marketing Level," in Ferrell, Brown, Lamb, eds., *Conceptual and Theoretical Developments in Marketing,* p. 679.

129. G. Zaltman, K. LeMasters, M. Heffring, *Theory Construction in Marketing* (New York, John Wiley, 1982).

130. J. F. Gaski, "Nomic Necessity in Marketing Theory: The Issue of Counterfactual Conditionals," *Journal of the Academy of Marketing Science,* 13 (Winter–Spring, 1985), pp. 310–320.

131. R. Bartels, "Physics and the Metaphysics of Marketing," in Winter, Gardner, *Converse Symposium,* p. 31.

3

Two-Dimensional Market Heterogeneity: Its Implications for Market Behavior, Market Strategy, and the Function of Markets

> [The] homogeneous market has no counterpart in the real world. It is only a convenient fiction adopted by economists who want to think about the economic problem of price rather than the marketing specialist who wants to think about the marketing problem of information.
>
> WROE ALDERSON[1]

The first half of this chapter shows that the idea of two-dimensional market heterogeneity is a useful heuristic for organizing, distinguishing, and understanding fundamental types of market behavior. These types can be viewed descriptively as forms of market behavior that, combined with price competition, provide a fuller and more balanced view of actual market behavior. They also can be viewed in a normative sense as strategy types. In this case the contribution of the idea of two-dimensional market heterogeneity lies in the identification of the fundamental demand-based behavioral sources of strategic advantage and the clarification of differences, complementarities, and possible incompatibilities of the strategies based on these sources of advantage.

The second half of the chapter deals with the implications of the two-dimensional structure of market heterogeneity for the function of markets in general. In particular, two-dimensional market heterogeneity is seen to imply a critical role for information. The emergence of this role in turn leads to the identification of a further source of strategic advantage: the exploitation of the inadequacy of market information and, especially, the reduction of customer risk.

TWO-DIMENSIONAL MARKET BEHAVIOR

The Fundamental Forms of Strategic Market Behavior

Two-dimensional market heterogeneity, as explained in Chapter 1, implies the potential for four basic forms of market behavior. In particular, behavior can exploit one dimension of heterogeneity, the other dimension, both, or neither. These combinations define four fundamental forms of market behavior, each of which can be distinguished by its marketing approach as well as by its underlying economic logic. The four forms of market behavior, how they are related and distinguished, their economic logic, and marketing approach are summarized in the three matrices in Chapter 1. As illustrated, different types of market behavior are distinguished by the dimensions of market heterogeneity they exploit. A direct comparison of each type of behavior with the other types further clarifies how they are related and distinguished.

The simplest form of strategic market behavior is price competition. Price-competition behavior takes advantage of neither the difference in customers nor the potential differences in products. It involves neither the disaggregation of demand nor the "kinking" of demand. Its appeal is directed to the general customer base and rests solely on price.

The role of market strategy in this form of behavior is the smallest of the four behavior forms because the effectiveness of price competition lies, primarily, in cost advantage: a supply-side source of superior performance. Because cost advantage depends largely on productive efficiency, the manufacturing function's role relative to marketing function's role is greatest in this form of behavior. The effectiveness of price competition in creating competitive advantage depends on marketing strategy only to the extent that the marketing function may have a role in improving the sale of a commodity through an appeal to low price and thereby help promote the efficiency advantages associated with scale economies, scope economies, and experience curve effects.

Product augmentation takes advantage of the potential heterogeneity in products but not in customer preferences. It is thus related to price competition in that it does not exploit customer heterogeneity and in that it is aimed at the general market as opposed to market segments. It is distinguished from price competition by the fact that it does exploit one dimension of heterogeneity: product heterogeneity. As discussed earlier, the economic logic of product augmentation lies in the concept of turning or kinking the demand curve so as to reduce customer price sensitivity and product substitutability, thereby facilitating the possibility of premium pricing, customer loyalty, and greater stability.

Because the creation of product augmentation may be dependent on the manufacturing function, manufacturing is important; but productive efficiency must be balanced with effectiveness in producing product augmentations that better satisfy customers. While product augmentation does not require the segmentation of the customer base—identification of subgroups or clusters of customer wants—it does imply a need to understand, and appeal to, the universal product needs of customers. Marketing's role is therefore greater than that implied in price competition.

Price segmentation takes advantage of the heterogeneity of customers but not products. It is similar to price competition in that it does not exploit the potential heterogeneity of products and in that it emphasizes price. It is distinguished from price competition by the fact that it does exploit customer heterogeneity in the creation of strategic advantage. As discussed earlier, the economic logic of price segmentation lies in the concept of disaggregating the demand curve, which is requisite to the extraction of consumer surplus and thus revenue in excess of that associated with price competition. As noted, legal restrictions may effectively inhibit an explicitly discriminating price policy. However, marketing product policies such as tie-in sales, metering, or bundling allow the firm to extract some consumer surplus without ostensibly engaging in discriminatory pricing among customers.

Price segmentation is related to product augmentation insofar as they both exploit a dimension of market heterogeneity. The two are distinguished in that price segmentation does not exploit product heterogeneity. The two are also related, along with price competition, in that they all attempt to appeal to the total market. Although price segmentation implies the segmentation of the customer base, price segmentation also implies an appeal to the whole market, for extraction of consumer surplus is dependent on the ability to appeal to as many price segmented groups as possible. Price segmentation is thus aimed at total demand, even though it involves the disaggregation of demand. In other words, price segmentation aims at "tapping" the total demand curve at as many levels as possible.

Product segmentation necessarily takes advantage of both customer heterogeneity and product heterogeneity. It is most unlike price competition because it requires the use of both dimensions of market heterogeneity. It is related to product augmentation in that it uses the potential heterogeneity of products and it is rooted in the economic logic of turning the demand curve. It is distinguished from product augmentation because it also takes advantage of heterogeneity as to customer preferences. Product segmentation is related to price segmentation in that it exploits customer heterogeneity and is rooted in the economic logic of the disaggregation of demand. It is distinguished from price segmentation in that it exploits product heterogeneity and the potential for genuine product augmentation and does not focus on price.

Product segmentation is not simply the use of both product heterogeneity and customer heterogeneity in one type of behavior. Bundling activity exploits both these dimensions of heterogeneity (as is shown later) but does not constitute product segmentation. Product segmentation is more complex than this because it is the exploitation of one dimension of market heterogeneity by means of the other. Customers are segmented on the basis of a product augmentation. And the product is augmented on the basis of the needs of a customer segment. Product segmentation is product-based segmentation or, in other words, segment-focused augmentation. Product segmentation is therefore further distinguished from product augmentation and price segmentation by the fact that it is a segment-focused strategy rather than a general market strategy.

To avoid ambiguity it is important to clarify the subdimensions of the general dimension of customer heterogeneity: customer heterogeneity as to price sensitivity and customer heterogeneity as to product preferences. Price segmentation exploits customer heterogeneity as to price sensitivities. Product segmentation exploits customer heterogeneity as to product preferences. Product segmentation also exploits product heterogeneity; price segmentation does not.

Product augmentation does exploit product heterogeneity but does not exploit customer heterogeneity as to product preferences. However, product augmentation as well as product segmentation may exploit customer heterogeneity as to price sensitivities if a product action is coupled with a price action. In particular, product augmentation necessarily involves a broad-based appeal from the customer-preference perspective (by appealing to a universal benefit) and simultaneously *may* involve an appeal to a segment of the market from a customer price sensitivity perspective.[2] Table 3.1 summarizes these distinctions.

Product augmentation and product segmentation will involve the exploitation of reservation price heterogeneity only if they are accompanied by a price change of sufficient magnitude in the same direction (positive). Conversely, product augmentation and product segmentation do not involve exploiting reservation price heterogeneity if they are not accompanied by higher prices. Both types of augmenting behavior necessarily imply only product behavior and not price behavior. Indeed, general or segment-specific augmentation does not even necessarily imply a higher product *cost,* as in the case of an innovation that may actually lower cost. Moreover, in addition to receiving a higher price an augmentor may want to expand or protect market share. In such cases, product activity is not necessarily accompanied by price activity that exploits reservation price differences.

Product segmentation, price segmentation, and product augmentation are all similar in that they imply an emphasis on understanding demand, the customer, and the value he places on a product. Product augmentation attempts to attract customers by improving the net benefits all customers associate with a product; price segmentation entails distinguishing customers on the basis of the value (the reservation price) they place on a product; and product segmentation attempts to add value through a more precise satisfaction of customer needs with augmentations that have a segment-specific appeal. All these market behaviors are fundamentally different from price competition, which relies exclusively on reducing a firm's costs per unit of output as a means to competitive advantage.

A comprehensive general framework of market behavior and strategy needs to balance the analysis of firm activity in terms of the degree of price competition with the three other approaches to understanding behavior: the exploitation of potential product heterogeneity by appealing to a universal benefit (product augmentation), the exploitation of both product and customer heterogeneity by appealing to a segment-specific benefit (product segmentation), or the exploitation of customer heterogeneity through various pricing techniques (price segmentation). The varied emphasis of different views on market behavior can be crudely summarized in terms of the attention given to each of the four behavior types. As a theoretical extreme,

Table 3.1 Type of Differences Exploited for Each Behavior Category

	Universally Appealing Product Differences	Segment-Specific Preference Differences	Customer Price Sensitivity Differences
Product augmentation	yes	no	maybe
Product segmentation	yes	yes	maybe
Price segmentation	no	no	yes

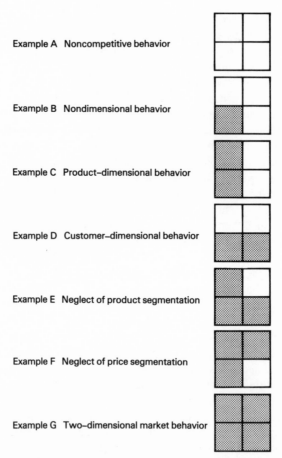

Example A   Noncompetitive behavior

Example B   Nondimensional behavior

Example C   Product–dimensional behavior

Example D   Customer–dimensional behavior

Example E   Neglect of product segmentation

Example F   Neglect of price segmentation

Example G   Two–dimensional market behavior

Figure 3.1. Perspectives on market behavior.

the behavior of a pure monopolist might be thought of as entailing no strategic behavior whatsoever in that competition does not exist (Example A of Figure 3.1). Contemporary economic models of behavior rooted in equilibrium price theory view competitive behavior as ultimately price competitive. This perspective on behavior is here termed "nondimensional" behavior because it neglects both dimensions of heterogeneity (Example B).

The monopolistic-competition model, the "workable-competition"[3] model, and the "complete-competition"[4] model, although all different, view competition primarily in terms of "product-dimensional" behavior, which includes both price competition and product augmentation in some limited sense (Example C). "Product dimensional" is the term applied here because it admits of product heterogeneity but neglects customer heterogeneity. Price-discrimination models are here characterized as "customer-dimensional" perspectives on behavior because, although the customer dimension of heterogeneity is used, product dimensionality is neglected (Example D).

If economic perspectives on behavior are considered in total, then three forms of behavior are emphasized but the intersection of product heterogeneity and customer heterogeneity—product segmentation—is neglected (Example E). When various separate pockets of marketing thought are considered in total, substantial attention is placed on augmentation, market segmentation, as well as price-cutting tactics. But there is less attention given to price segmentation (Example F) for reasons discussed in the previous chapter. This study calls for a more complete and balanced framework for strategic market behavior, the formulation of which lies in the two-dimensional nature of market heterogeneity (Example G).

The Complementarity of Product Augmentation and Price Segmentation

Price segmentation, unlike the other behavior forms, does not necessarily imply a commitment to either gaining a cost advantage (as does price competition) or a commitment to investing in effective augmenting behavior (as does product augmentation and product segmentation). Varying the price of a product requires no specialization of production resources and relatively little, if any, commitment of marketing resources. Price segmentation may not even require the explicit identification of customer segments because, as explained in Chapter 2, several sophisticated forms of this behavior result in the extraction of consumer surplus automatically through the buying arrangement (i.e., bundling).

Because price segmentation focuses on neither cost reduction nor augmentation along product dimensions and therefore does not require a diversion of funds to cost reduction and/or product enhancement, it tends to be less demanding in terms of scarce organizational resources than other behavior forms. Therefore, the point at which it begins to compromise another strategy, as a result of the need to trade off scarce organizational resources between strategies, may occur only after a much higher level of resource commitment. It thus implies less incompatibility if used jointly with another strategy than do other strategy mixes.

Further, price segmentation and product augmentation can be mutually reinforcing. To the extent that price can signal product quality differences, price segmentation may reinforce product differentiation. To the extent that differences in products can be used to justify differences in price, differentiation may reinforce price segmentation. Effective price segmentation depends, in part, on the ability to inhibit the substitution of the products sold by rival firms for the product sold by the discriminating firm. The presence of competitive product substitutes reduces the ability to implement a segmented pricing program. Because product augmentation reduces price sensitivity, cross-price elasticity, and thus product substitution, it sets the stage for price segmentation. Indeed it is such reduced price sensitivity that permits price segmentation despite the absence of pure monopoly. One general product augmentation, such as the superior reputation of IBM, may create a level of insulation from product substitution sufficient to carry out an effective price segmentation program.

There is even further complementarity between product augmentation and price segmentation when the specific marketing practice of product bundling is considered. Product bundling can facilitate both the exploitation of product heterogeneity and customer heterogeneity. That product bundling can be a means to price segmentation was discussed in Chapter 2 and is developed and revealed to be applicable to IBM's strategy in Chapter 5. Chapter 4 will show that bundling is also a vehicle to product augmentation in that it is a way for a seller to offer customers a low-risk "total solution" rather than merely a physical product. When this marketing technique is used, therefore, the implementation of two strategic objectives is accomplished in one activity. The issue of resource allocation between strategies becomes moot, and the cost of each strategy is reduced because the cost of one activity, bundling, is shared between the strategies.

Potential Incompatibilities within Strategy Mixes

There is a great paucity of empirical research examining potential incompatibilities among different approaches to market strategy and certainly no conclusive evidence either confirming or refuting hypotheses as to potential tradeoffs among strategies. Conventional business strategy literature offers deductive hypotheses concerning potential tradeoffs. In particular, it argues that *in general* two types of tradeoffs exist: a tradeoff between cost-based strategies and customer-value-based strategies, and a tradeoff between focused market-segment strategies and unfocused general-market strategies. More specifically, it is argued that while many firms can successfully pursue, simultaneously, both cost and value strategies or both focused and general strategies, there comes a point where, under sufficient competitive intensity, a firm should choose between one strategy or the other strategy. At this point one strategy is pursued only at the expense of the other. A firm that fails to make this choice in a sufficiently competitive industry will find itself "stuck in the middle" between firms that have emphasized either one strategy or the other (achieved superior cost or market focus on the one hand, and customer value or general market appeal on the other hand).

The arguments supporting this view range from those pointing to the scarcity of organizational resources and specifically tradeoffs between the functions of marketing and production[5] to those pointing to a variety of organizational variables, apart from resource scarcity, that imply a tradeoff, such as the organizational variables of culture, management style, structure, human resources, and systems. (This latter set of factors is discussed in the Conclusion, Chapter 7.) An articulate spokesman for the presence of a tradeoff is Porter, who argues that at some point further implementation of one strategy implies a sacrifice of the other strategy because cost (due to finite organizational resources) becomes a significant variable:

The benefits of optimizing the firm's strategy for a particular target segment cannot be gained if a firm is simultaneously serving a broad (market) . . . this implies a U-shaped relationship between profitability and market share.

Achieving cost leadership and differentiation are also usually inconsistent, because differentiation is usually costly. To be unique and command a price premium, a differentiator deliberately elevates costs. . . . Conversely, cost leadership often requires a firm to forego some differentiation by standardizing its product, reducing marketing overhead, and the like. . . . When faced with capable competitors also striving for cost leadership, a firm will ultimately reach the point where further cost reduction requires a sacrifice in differentiation. It is at this point that the . . . strategies become inconsistent and a firm must make a choice. A firm that engages in each . . . strategy but fails to achieve them is stuck in the middle.[6]

The idea of "stuck-in-the-middle" refers to the failure to generate adequate commitment to one strategy type in the first place as well as the failure to sustain that commitment. It is this later failure that is the danger of a "strategy shift." For example, a segment-focused competitor, lured by the larger potential revenue base of a general market, may attempt to alter its product so as to give it a broader appeal. But if broadening its appeal to the general market implies a reduction in its appeal to its original market segment, than the strategy may backfire and the firm may find itself with a product insufficiently customized to its customer segment to protect it from competitive substitutes and insufficiently low priced or augmented to make it a viable competitor in the general market. Chapter 6 shows this is exactly what happened to Control Data Corporation in the early years of the supercomputer industry.

Given the smaller potential revenue base targeted by this behavior and the costs associated with it, why is it undertaken at all? One major reason is it may be the only viable means of competing. Levitt, for instance, argues that general market augmentation "creates survival problems for the small members of industries dominated by large firms. It is to be expected, therefore, that survival for the small firm will increasingly take the form of carving out highly specialized niches in markets that the large, think-big firms cannot reach."[7] For example, the highly effective general product augmentation of IBM in the general market (discussed in Chapter 4) may present an insurmountable barrier to entry in the general market. Rather than compete head to head in general markets with established firms having greater resources to implement their market strategy, other firms can, in effect, redefine the market for their purposes and specialize their effort to gain strategic advantage with respect to that niche only. The fault-tolerant computer and supercomputer niches discussed in Chapter 6 are cases in point.

Apart from case study confirmations of incompatibility between general and segment-focused strategies, such as that discussed in Chapter 6, this author could not find empirical work directed at this strategy question. Further, there is little generic empirical research into the potential tradeoff between cost and value-based strategies.[8] Moreover, what work has been done on this topic uses product quality as a proxy variable for superior customer value.

Phillips, Chang, and Buzzell summarize the empirically based argument for a quality/cost strategy tradeoff:

Higher quality usually requires the use of more expensive components, less standardized production processes, and the adoption of other manufacturing and management techniques incompatible with achieving low costs. Furthermore, achieving a high quality position may require higher expenditures in other areas beyond the direct costs of manufacturing and distribution. Higher advertising and promotion expenditures may be necessary to convey a quality position to customers; increased sales force spending may be needed to support the higher level of customer services that may accompany higher quality products; and a heightened emphasis on product innovation may be necessary to sustain a quality position. Finally, achieving such a position may require a perception of exclusivity, which is incompatible with the high relative market share needed for a volume based cost leadership strategy.[9]

Supporting this view, Farris and Reibstein[10] show that promotional activity associated with a customer-value strategy requires a higher level of expenditure than otherwise to convey a quality image to customers.

Conflicting with this view, however, is notable evidence that suggests product-quality strategies and cost-based strategies may, in fact, be mutually reinforcing. This evidence takes two forms. The first type suggests that, with certain types of production processes and with certain approaches to production operations management, factors that contribute to lower cost also contribute simultaneously to higher quality.[11] The second type of evidence suggests that higher quality can indirectly, through its impact on market share, improve a firm's cost position as well.

Supporting the first line of reasoning, for instance, Fine argues that because production workers spend more time and care on work activities that produce high-quality products than activities producing low cost products, cost declines more rapidly with accumulated experience for high-quality products.[12] Additional care spurs the identification and correction of defects in the production system that are otherwise missed by less careful work associated with low-cost production. The proposed phenomenon of a "quality-based learning curve" implies a mutually beneficial relationship between cost and quality-based strategies. Wheelwright[13] and Fine[14] argue that the empirically established advantages of Japanese manufacturers in both cost and quality cannot be entirely explained in terms of disparities in wage rates, capital investment, or factory automation, but rather must also be accounted for in terms of a strategic approach to operations that improves both cost position and quality simultaneously.

Concerning the second type of contrary evidence, Buzzell and Wiersema,[15] as well as Flaherty,[16] argue that product quality indirectly improves cost position through its positive impact on market share. This line of reasoning rests on the phenomenon of the traditional cost-based experience curve coupled with an empirically established linkage between high product quality and higher relative market share. In particular, higher quality leads to higher share in certain industries, which in turn leads to a better cost position due to experience curve effects such as economies of scale and learning curves. Moreover, Wensley and Robin[17] argue experience effects apply not only to costs of production but to all value-added activities. Therefore, given the positive impact of quality on share, higher quality and thus greater share do not necessarily imply higher expenditure on promotional activity, sales force activity, and other marketing functions.

Phillips, Chang, and Buzzell tested the relationship between product quality and cost position using the PIMS[18] data base. Their findings failed to confirm the view that a high relative quality position is incompatible with achieving a low relative cost position; but they also found that in certain industries (such as the "components businesses") "higher product quality does require a tradeoff in direct cost position."[19]

The major drawback in applying any quality/cost studies to the hypothesis of a tradeoff between customer-value and cost-based strategies is the inadequacy of product quality as a proxy for perceived customer value. Even if it were conclusively shown that, generally, quality is not incompatible with low cost, other variables affecting perceived value (such as service) still may be incompatible. In

Table 3.2 Potential Internal Inconsistencies of Strategy Combinations*

Possible Combinations				Potential Tradeoff		
Price compe- tition	Product augmen- tation	Product segmen- tation	Price segmen- tation	Cost-Based versus customer- value based	Focused versus general market	Either tradeoff
✓	✓			Yes	No	Yes
✓		✓		Yes	Yes	Yes
✓			✓	Yes	No	Yes
	✓	✓		No	Yes	Yes
	✓		✓	No	No	No†
		✓	✓	No	Yes	Yes

*Check marks indicate which strategies form combinations.
†Only combination (product augmentation and price segmentation) that implies no internal inconsistencies.

fact, effective customer value competitors, such as IBM, may achieve that superior value position for reasons having little to do with actual physical product quality advantages, thereby rendering quality arguments, either way, moot.

The cost/value strategy compatibility debate notwithstanding, the *potential* incompatibilities among the four behavior forms developed in this thesis are summarized in Table 3.2. All the possible combinations, with the exception of the product augmentation/price segmentation combination, imply a potential internal inconsistency due to either the possibility of a cost/value tradeoff or a general market/segment-focused strategy incompatibility. Interestingly, while each of the four strategy types is individually present in the mainframe computer industry, Part II of this thesis reveals that the only strategy *mix* observed in the mainframe market is the same strategy combination that in this chapter is shown to be the only one free of potential internal inconsistencies for theoretical reasons.

FROM MARKET PRICE TO MARKET INFORMATION

Not only does two-dimensional market heterogeneity lead to a more balanced perspective on market behavior, but, as outlined in Chapter 1, it also leads to a more balanced perspective on the very function of markets. If the fundamentally heterogeneous nature of markets is acknowledged and its two-dimensional structure clarified, the view emerges of the function of markets as that of matching: matching heterogeneous products with heterogeneous customers toward the goal of market effectiveness, not just production efficiency. Two-dimensional market heterogeneity, and its implication for the important role of matching in the function of markets, in turn, implies a need to readjust the emphasis on price as the exclusive mechanism in the function of markets to its position as a subset of market information, the prime driver of the matching process.

As a study of the price system, conventional economic models of competition are perhaps unparalleled in their elegance among theories in social studies. But to use these models of the price system as a comprehensive framework of market behavior is to misunderstand the very subject of these models and to neglect the significance of information. As Demsetz argues:

> If the perfect competition model is seen as a tool for understanding the price system, and not for understanding competition, it represents a natural evolution from and vital capstone to the central interests of the classical writers. Part of the difficulties with competition come from our attempt to use the perfect competition model for a purpose for which it is not ideally suited—for analyzing competitive activity. . . . It is the price system the model explicates, not competitive activity. Competitive activity itself is difficult to comprehend through a model that assumes away transaction and information costs.
>
> The special notion of competition relied upon by the perfect decentralization model makes that model a poor vehicle for understanding a wide variety of competitive tactics and institutions that are adopted precisely to accommodate to time, uncertainty, and the cost of transacting. Particular marketing practices, such as tie-in sales . . . are difficult to explain with a model that assumes away their cause.[20]

Consistent with Demsetz's observation, this thesis suggests that the perspective on market behavior that places overwhelming emphasis on price is a misdirected application of economic thought as a comprehensive explanation of market behavior rather than as an explanation of one component of market behavior. The most significant shortcoming of using economic models in this way is the resulting neglect of the implications of heterogeneity for behavior and of the role of information in the functioning of markets.

Traditionally marketing theory views price as one of many elements in a "marketing mix," although the precise importance of price in that mix is debated in the marketing literature.[21] The contribution of this literature to the argument of this book lies not in the degree of importance placed on price but on the idea that price is just one of many important variables in the purchasing decision process. Price is important because it contributes to the consumer's evaluation of the net benefits of a product. The specific role of price is as a cost in the total bundle of costs and benefits the consumer purchases. In contrast to this, customer choice under homogeneous conditions is necessarily based solely on price. There is no other purchase criterion. It is heterogeneity that gives rise to the need to consider customers' perceptions of the net benefits of the products.

This study does not attempt to diminish the special role of price in market behavior; on the contrary, price is a uniquely important piece of market information in that it not only is a cost to the consumer but revenue to the seller and may serve as a common denominator for product comparison. Further, under the theoretical market conditions of homogeneity and complete market information, the purchasing process does reduce to only a response to price.

These conditions, however, should be treated as theoretical reference points only; actual conditions approaching these (unchanging markets characterized by ultrasophisticated customers and stagnant product technology) are rare. As Alderson argues:

> The emergence of relatively homogeneous conditions at certain stages of the competitive process [should be] treated as a tendency which can be functionally useful for some aspects of marketing operations rather than as an essential aspect of effective competition. . . . This is in sharp contrast to the approach that starts with a competitive ideal and finds itself obliged to

reduce behaviour systems such as firms to . . . abstract entities because the going concern of real life does not fit the pattern. This is [not] to detract from the great achievement of economists in developing their deductive analytical apparatus to the point where it approximates the view of competition which is obtained more directly by making a fresh start. . . . It may [be] that the differences of viewpoints are appropriate to differences in the problems which each discipline sets out to study.[22]

The economic perspective on markets and market behavior is ultimately limited by the tools it uses to understand them. Economic analysis, rooted in the demand curve framework, is built upon certain assumptions. The most important of these for our purposes are the assumptions of homogeneity and complete information. Given these starting points, economics erects an elegant analytic framework. In order to make this framework more relevant to actual market behavior, it is amended by describing, for instance, how a demand curve changes if the assumption of product homogeneity is relaxed. But the starting point, the "ideal type," the "first-best" case, remains the same. Because of this conceptual foundation, the phenomenon of two-dimensional heterogeneity and inadequate market information are not readily expressed in terms of a demand curve. The essential features of a more robust analytic apparatus for understanding market behavior require the removal of homogeneity as a starting assumption and its replacement with two-dimensional heterogeneity as an inherent feature of the pure model of behavior, not a violation of the pure model. This results in a more balanced view of markets to include their function of "matching" heterogeneous products with heterogeneous customers, toward the end of not only efficiency, but also effectiveness, and through not just price mechanism but the superordinate vehicle of market information.

THE TWO SIDES OF HETEROGENEOUS MARKET STRATEGY

The concept of the function of markets as a matching process between customer wants and product benefits has implications for market strategy. Up to this point in this thesis, explanations of product augmentation, product segmentation, and price segmentation have implicitly assumed that customer wants are a given. For example, product augmentation only makes sense in the context of some universal customer want that products are not fully satisfying and toward which product augmentation efforts are directed. Price segmentation only makes sense in the context of preexisting differences in customer price sensitivity. Finally, product segmentation only makes sense under even stronger assumptions about the structure of customer heterogeneity. Product segmentation specifically requires the "segmentability" of customers or, in other words, the presence of customer differences that are insignificantly small within a segment of customers and significantly large between this segment and the rest of the customer base.

What the notion of matching reveals is that there are two sides to each behavior type. In particular it reveals the important potential for altering the previously assumed fixed nature of customer wants as a vehicle for achieving product augmentation, product segmentation, or price segmentation.

Product augmentation therefore must be thought of as a relative phenomenon:

that is, it is not simply change along the product dimension but change in products relative to those product attributes desired by customers. The notion of product augmentation as product change relative to customer wants implies that it can be accomplished by not only changing the perceived product but also by changing customer wants to fit existing differential product advantages. More specifically, product augmentation can take three possible variations: changing the product or the perception of the product to fit customer wants (Chamberlinian economic logic), changing customer wants to fit some existing product difference, or some combination of both. The underlying market condition required for all three variations is that product offerings do not fit the nature of customer wants. The difference among these variations is the side (or sides) of the matching equation that is changed to improve the product/customer fit: the product side or the customer side or both. Chapter 4 shows that these distinctions are helpful in understanding the strategic logic underlying IBM's marketing policies of product bundling and customer account management, which focus on different sides of the matching equation.

The same distinctions hold for approaches to product segmentation. A change in the customer side of the matching equation, in the case of product segmentation, implies altering the nature of customer wants *of a segment of the customer base* so as to bring this segment into a closer fit with a unique product benefit presently available in a product offering. Product segmentation, like product augmentation, can hence take any of three variations: changes in segment-specific, perceived product benefits; alterations in the wants of a specific customer segment; or a combination of both. Consistent with the basic distinction between product augmentation and product segmentation, product *augmentation* through altering customer wants does not necessarily imply heterogeneity along the relevant customer dimension (either before or after the alteration), while product *segmentation* through altering customer wants does imply heterogeneity along the relevant dimension (at least after the alteration).

Finally, the concept of *price* segmentation is also extended when the possibility of altering customer wants is considered. Price segmentation, as described up to this point, involves the exploitation of *existing* customer heterogeneity and, specifically, existing differences in customer price sensitivities. However, the possibility of changing the nature of customer heterogeneity, which the customer side of the matching process implies, reveals another side to price discrimination not explicitly considered in the Pigouvian economic logic of price discrimination or the modern price segmentation frameworks on which logic they are based. Specifically, alteration of the nature of customer heterogeneity can take the form of changing the degree of a variability in customer price sensitivity (from, for instance, moderate price-sensitivity differences to more extreme differences in price sensitivity) thereby improving the potential gains from a segmented pricing policy. Given the presence of a tiered price structure, changing customer heterogeneity in this way increases the seller gains from price segmentation. In the absence of a tiered price policy, it creates and/or improves the conditions necessary for price segmentation.

The customer side of price segmentation is therefore the exaggeration of differences in price sensitivities; the other side of price segmentation is the actual pricing of the product at different levels in order exploit the existing differences in price sensitivities. Chapter 5 shows clearly that IBM's price segmentation not only includes differential pricing that exploits the existing heterogeneity in customer price sensitivities, but also entails the manipulation of customer price sensitivity differences by altering computer interfaces and promoting customer "lock-in" in order to amplify the effects of a tiered price strategy.

Changing a product or changing customer wants can both result in an alteration of the demand function (i.e., the relationship between the quantity demanded and the independent variable of price) as opposed to merely changing the independent variables in this function. The means of altering this functional relationship between price and quantity demanded (price sensitivity) cannot be fully understood in terms of product differentiation; rather, it requires the fuller idea of matching product benefits and customer wants. Specifically, it requires recognition of the two sides of product augmentation: changes in the perceived product and/or changes in customer wants to better match some existing product difference. This implies that the distinction made by Dickson and Ginter between a "product-differentiation strategy" and a "demand-function modification strategy"[23] as discussed in Chapter 2 is ill-conceived, or at least poorly phrased, insofar as product change is a means to demand function modification: a goal these authors associated with the alteration of customer wants only. Once the matching function of markets is clarified, it becomes clear that both approaches are flipsides of the same coin (improving the product/customer match) and that both approaches rest on the economic logic of changing the functional relationship between price and quantity demanded.

The demand-curve framework is of limited utility for illustrating the distinction between changing the perceived product and altering customer wants as it applies to product augmentation and product segmentation because the framework assumes away differences in nonprice product dimensions critical to these behavior forms. More helpful than the demand-curve framework in distinguishing changes in the perceived product from changes in customer wants is the product-preference-map framework based on the multiattribute model of customer demand[24] associated with the product policy area of the marketing discipline and related to the economic literature on optimal product variety.[25]

Preference maps can be used to represent the conditions of product heterogeneity, customer heterogeneity, as well as customer segmentability. Customer wants or preferences are represented as ideal points in a preference space. Products are also represented as points in the space that reflect customers' perceptions of products. Product heterogeneity is represented by differences in the position of the product point of each seller. Customer heterogeneity is represented by differences in the position of customer preference points. Customer *segmentability* can be thought of as the presence of different clusters of preference points within the preference space.

Although typically used to generate specific empirically researched market segmentation programs, preference maps are used in this thesis to illustrate and clarify the logic of the theoretical distinctions (developed above) between alterations in customer wants and alterations in perceived products. In Example A of Figure 3.2, the horizontal (x) dimension represents a product attribute that tends to have univer-

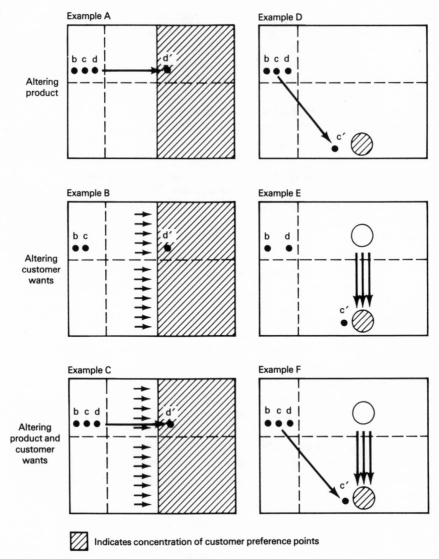

Indicates concentration of customer preference points

Figure 3.2. Two approaches to improving product/customer match (illustrated with preference space diagrams).

sal appeal such as customer service. Hence, customer ideal points or preferences, although evenly distributed along the vertical (y) dimension, tend to cluster toward one extreme along the x axis (represented by the shaded area).

Product augmentation is facilitated by altering the perceived product along this dimension (for instance by actually improving service or enhancing the firm's service image) so as to bring the product closer to customer preferences. This is diagrammatically portrayed as the movement of product d, by the firm making this product, away from the products of other firms along the x dimension (to a higher service level) closer to the concentration of customer preferences (see Example A).[26] Alternatively, given a less clear universal preference for the "x" dimension attribute (service), or in other words a less concentrated and more distributed set of preferences, product augmentation could be facilitated by altering customer wants to better match the existing product advantage of the firm selling product d' (see Example B). For example, instead of improving perceived service, this approach assumes superior service and attempts to intensify the customer's desire for superior service by, for instance, making the customer more aware of the risk to their business of product failures.

In summary, where an existing product advantage exists, product augmentation can be facilitated by molding customer wants around existing product advantages. In this way, the firm is exploiting *product* heterogeneity by altering customer wants. Alternatively, where product heterogeneity does not already exist, a firm can exploit the potential for product heterogeneity by altering the product directly. Further, there is no reason why these approaches would be mutually exclusive, and thus, as Example C suggests, the matching process can be approached from both the product side and the customer side simultaneously (for instance, by improving both the level of perceived service and increasing awareness of the importance of service to the customer).

Similarly, given heterogeneity in customer wants and the segmentability of customers (a clustering of preference points), product segmentation is facilitated by altering the perceived product along one or more dimensions that appeal to the particular want that forms the basis of the distinct cluster. In other words, the product is targeted to distinctive customer segments. This is illustrated by the product movement of c to c' (Example D). Conversely, given a differential product c', the firm can attempt to modify wants of some customers so as to form a customer niche in which the firm's product has a clear advantage (Example E). In the first case, both product heterogeneity and customer heterogeneity are exploited directly through product actions. In the second case, both dimensions of heterogeneity are exploited by altering customer wants so as to better fit with the firm's unique product advantage. As with general market augmentation, there is no reason why both approaches to product segmentation cannot be used together (see Example F).

INFORMATION INADEQUACY

As explained, the significance of market information for the function of markets arises from two-dimensional market heterogeneity and specifically the resultant role of information in matching heterogeneous products with heterogeneous customers. The significance of market information for market strategy lies in its inadequacy and, in particular, the inadequacy of information about products among customers needed to make purchasing decisions. The effects of this inadequacy were outlined in Chapter 1. Information inadequacy diminishes the ability of the buyer to select among products (low buyer sophistication) and particularly the ability to reduce products to some "implicit price" or common price-performance measure facilitating product comparisons. Inadequate market information also increases the customer's receptivity to information signals from the seller that substitute for direct information. Most importantly, inadequate market information leads to the phenomenon of perceived customer purchase risk. The *strategic* import of market information rests in the ability of the seller to take advantage of these effects of information inadequacy.[27]

Asymmetric Information

The importance of information inadequacy in market behavior begins with the disconnecting of the direct link between product substitutability and price sensitivity.[28] In the conventional economic view, price elasticity is a direct function of product substitutability (among other factors). Perfect substitutability implies perfect price sensitivity. Product differentiation implies diminished substitutability and thus diminished price sensitivity. However, recent economic work incorporating market information into market behavior frameworks points out that, irrespective of the level of objective substitutability, price sensitivity may be low due to inadequate information. Information adequacy serves as an intermediate variable between actual substitutability and price sensitivity. In other words, in terminology not used in the recent economic contributions, it is the *perception* of substitutability as affected by the adequacy of market information, not substitutability itself, that determines price sensitivity.

The impact of the variable of inadequate market information on price sensitivity is important because it has implications for the effectiveness, and thus the appropriateness, of different market strategies. Under perfect information conditions, premium product quality is readily perceived by the customer. Sellers can extract price premiums that reflect the greater utility of higher quality products because customers can reliably assess quality and thus readily compare products on the basis of price. Price behavior is therefore an effective competitive mode and price premiums reflect the actual utility of superior quality and no other factors.

Under "asymmetric information"[29] conditions and, in particular, inferior customer information relative to seller information, the seller is unable to extract a premium commensurate with the higher quality level without some mechanism that signals the premium quality to the buyer. This is so because inadequate information among buyers inhibits their ability to make reliable comparisons among competitive products. As a result, price is less important in determining buying decisions and "price competition [is] less effective in markets with asymmetric information."[30] Conversely, product dimensional behavior that signals information to the customer and/or reduces purchase risk that arises from inadequate information becomes even more important than it would be under more balanced information conditions.

The price premium required by the premium-quality seller in markets with information asymmetry is higher than it would be under symmetric conditions because, as Klein and Leffer point out, the premium product "must not only compensate the firm for the increased average production costs incurred when [higher quality] is produced, but most also yield a normal rate of return on the foregone gains from exploiting consumers' ignorance."[31] This is so because, as Nagle explains, "a seller whose price just reflected the extra cost of high quality would have an incentive to reduce his quality (and production costs), thus earning extra profits until buyers learned about the reduction through disappointing purchases."[32] Therefore, as Shapiro concludes, "firms producing high quality items must be earning a flow of profits to prevent them from being tempted to cut quality."[33]

This point is important in explaining otherwise inexplicable price premiums. For instance, a major study of mainframe computer price differentials,[34] discussed in Chapter 4, found that even after performance quality differences were removed, IBM still received higher prices for its products. How can this difference be explained? One explanation offered by the authors was that the price premiums simply reflected the excess, unearned, and unfair profitability associated with monopoly power.

The economic logic discussed above suggests a possible alternative explanation. Price premiums, beyond that which are explainable in terms of higher actual utility to the customer, may reflect the opportunity cost to the seller of foregoing profits from a short-term strategy of reducing quality and exploiting temporary buyer ignorance, in favor of a long-term strategy of maintaining quality. In other words, price premiums in excess of performance differences are not necessarily the unfair extraction of unearned revenue stemming from monopoly power. Under conditions of asymmetric information, premiums over those indicated by utility differences may simply reflect the seller's opportunity cost of foregoing a "fly-by-night strategy" for a "faithful strategy."

Purchasers who prefer higher quality levels will be subject to the "price-quality" effect in markets with asymmetric information.[35] Given a price premium reflecting both quality and the opportunity cost of exploiting consumer ignorance as discussed, a seller is motivated to maintain the quality. At a lower premium, the seller is motivated to reduce quality and exploit consumer ignorance. If the purchaser recognizes this he may rely on price as an indicator of quality. The higher the price premium, the more credible the higher quality level; the lower the price premium, the more dubious the quality difference.

Price-competitive behavior therefore is not only ineffective in markets with asymetric information but can be positively detrimental to the position of effective product augmentors in that it can remove the credibility of perceived augmentation. A further implication of asymmetric information is therefore the potential incompatibility of price-competitive behavior and product-augmenting behavior. To summarize, asymmetric information conditions lead to the ineffectiveness of price competition in general, to create a disincentive for an existing augmentor to engage in price-competitive behavior, provide an alternative explanation for excess premiums other than "monopoly power," and intensify the important role of product-dimensional behavior, especially product-dimensional behavior that exploits the effects of inadequate information.

Information Acquisition

Like the economic framework of imbalanced information, the economic logic of "information acquisition costs" is also a helpful heuristic for thinking about strategy that exploits inadequate information. Nelson argues that the cost of acquiring information depends upon the nature of a product's characteristics.[36] This has two important implications for product policy. The first is that in industries where the nature of product characteristics imply high acquisition costs, there exists an environment characterized by price insensitivity and conducive to competition along nonprice, product dimensions. Second, the relationship between product dimensions and acquisition costs suggests a normative prescription for product policy formation. The success of a product differentiation policy in reducing price sensitivity and protecting a firm from price competition will, in part, depend upon the type of product dimensions along which the firm differentiates its products. In particular, augmentation along product dimensions where information acquisition is more costly is likely to be more effective in reducing price sensitivity because the buyer is less likely to reduce these augmentations to price/performance values that can be compared to the price/performance of competitive products.

Economists have categorized product attributes into three classes: experience attributes, search attributes, and credence attributes.[37] The basis for the distinction is how buyers learn about the attributes. Search attributes are those that the consumer can readily evaluate prior to purchase. Consumers learn about these attributes in the process of product search. In a world of perfect information all attributes would be search attributes. Purchase price is a good example of such an attribute. Experience attributes are those that the consumer can accurately evaluate only after the product is purchased. Consumers learn about these attributes only through knowledge of past experience with the product. In addition to these two categories associated with Nelson, Darby and Karni[38] identify a third attribute category: credence attributes. Credence attributes are those that consumers cannot confidently evaluate even after purchase. Here customers place a heavy reliance on product reputation and company image even for repeat purchases.

The importance of these categories lies in their relation to information acquisition costs. As the attributes that distinguish products move from search to experience to credence types, the cost of acquiring information to evaluate the products increases. For economic reasons, consumers will remain less able to evaluate and choose among products characterized by credence attributes. As Nagle explains:

As one moves from the search to the experience and on to the credence category, information about brands' differentiating attributes becomes more costly. The greater the cost of information, the less of it people try to obtain. Consequently, other things equal, consumers should choose to inform themselves about the differentiating characteristics of fewer brands, and inform themselves less completely, in categories with high costs of collecting such information. The fewer brands about which buyers are informed . . . the less price sensitive they will be to the price of any one brand. As a result, the effectiveness of price competition is reduced as is the ability of new competitors to enter the market with a strategy of low, penetration pricing.[39]

The implications for market strategy are straightforward. Industries categorized by search attributes for distinguishing products will be forced to compete on price. Such is the case in the airline industry. The product is distinguished on the basis of price, departure times, type of place, and airports used. Price competition is intense. Dishwashing liquid is chosen more on the basis of past experience. In this market, competition occurs more along experience-type attributes and other search attributes. Computer industry products are distinguished, in many cases, along attributes that are very difficult to learn about even after purchase. As a result, the importance of price in determining purchase decisions is likely to be lower in computer markets than in markets where information is less costly to obtain. (Chapter 4 shows that price, in fact, is one of the least important factors in computer product purchasing decisions.)

Information Signaling

Not only does information inadequacy (and specifically asymmetric information and high information acquisition costs) diminish the strategic import of price competition, but it also intensifies the importance of non-price-oriented market strategy aimed at providing signals to customers that substitute for direct information. The greater the inadequacy of market information, the greater the need for the customer to infer information and evaluate products on the basis of cues or signals, and hence the greater the potential role of cues in market strategy. Chapter 2 discussed how product augmentation can be facilitated by focusing on the benefit side and the cost side of the perceived net benefit equation. Information cues facilitate augmentation through their affect on the perception component of the perceived net benefit equation. In fact, the psychological construct of "verdical perception"[40] in which the perceiver adds information to that contained in the stimulus itself provided a starting point for work on information signaling.[41]

One of the first information signals studied was price as an indicator of product quality.[42] The idea that price is not just a datum in and of itself but also serves as a signal for further, nonprice product information has become accepted knowledge in marketing literature, although the conditions under which this relation is strongest are debated.[43] Part of the problem lies in the methodologically problematic nature of "single cue" studies (studies that look at only one cue, price, as a signal of product information) because, as Monroe argues "when price is the only cue available, subjects naturally associate price and quality."[44]

Multicue studies strongly suggest that seller size, success, reputation, and image also signal information about products.[45] Product information cues or signals that have been identified in the business literature specifically include: parent company identity, financial stability, time in business, installed base, customer list, market share, business size (revenue or units sold), attractiveness and size of facilities, and the visibility of the seller among the buyer's top management.[46] (Chapter 4 presents evidence that suggests such factors as stability and top management visibility are indeed important in determining purchasing decisions in the market for computer products.)

The concept of information signaling meshes well with economic perspectives on market behavior that incorporate inadequate information. Shapiro argues that it is, in fact, inadequate information that gives rise to the critical role of reputation as an information signal, a variable not considered in conventional models of market behavior:

The idea of reputation . . . the "goodwill" value of the firm's brand name or loyal customer patronage . . . makes sense only in an imperfect information world. If product attributes were perfectly observable prior to purchase, then previous production of high quality items would not enter into consumers' evaluations of a firm's product quality. In such cases a firm's decision to produce high quality items is a dynamic one: the benefits of doing so accrue in the future via the effect of building up a reputation. In this sense, reputation formation is a type of signaling activity.[47]

Just as the introduction of asymmetric information into market behavior models results in an alternative explanation for price premiums in excess of actual utility, the introduction of reputation results in a similar alternative explanation. Shapiro argues that not only does an excess premium reflect a "faithful strategy [of] forego-ing the opportunity to earn profits through quality reductions" as the asymmetric information model suggests, but the "premium also serves the function of compen-sating sellers for their investment in reputation."[48] Shapiro concludes:

Care must be taken in evaluating profit data . . . if reputation (goodwill) is not included in the set of assets a firm owns, the calculations of its rate of return will exceed the market rate of return. This is misleading, as would be the conclusion based upon it that the firm enjoyed some degree of market power.[49]

Customer Risk

The customer's perception of purchase risks arises from the inadequacy of market information. Perfect information as to both the outcome and consequences of a purchase decision precludes risk. The importance of customer risk for business strategy lies primarily in the possibility of reducing it as a form of product augmen-tation and, to a lesser extent, in the possibility of grouping customers according to risk categories. The reduction of perceived risk is a vehicle for exploiting the product dimension of market heterogeneity, while the segmentation of customers into groups according to risk types serves as a tool for taking advantage of customer heterogeneity.

Marketing research has suggested the prevalence of risk-averse purchasing behav-ior at both the organization level and the individual level.[50] The level of risk to an organization of a particular purchase is positively related to the size (financial worth) of the item purchased,[51] the complexity and novelty of the item purchased,[52] and the importance of the item purchased to the customer's business system.[53] (Chapter 4 shows that such purchasing conditions are present in the market for computer mainframe products.) The last category, the importance of the item pur-chased to a customer's business, is especially important in industrial markets where the products sold are factor inputs used for the production of other goods or ser-vices. As Howard explains:

In the industrial buy, unlike the consumer purchase, the risk can be literally a thousand times the invoice price. For example, unavailability of a $10 production tool can shut down an automated line producing tens of thousands of dollars worth of crankshafts a day. Again, a $10 tool, made incorrectly and installed undetected, can produce scrap parts worth thousands of dollars before the error is corrected. Delays in delivery of a piece of equipment costing a few hundred dollars can result in lost production and profits many times the value of the buy.[54]

To the extent that the goals and motivations of the individual (or individuals) within a company who actually make the purchase decision parallel the objectives of the firm, marketing behavior that reduces firm risk will be effective in attracting customer purchases. However, research suggests that "nonrational," "social," or individual-level goals not necessarily consistent with firm objectives play an important role in purchasing decisions. Insofar as such individual-level objectives do not reflect firm objectives, risk-reducing strategy must also address the perception of personal risk by the individual actually making a particular purchase decision for a firm.

The importance of individual-level risk in industrial purchasing behavior has been recently established in marketing literature. Webster, for example, found that industrial purchasing behavior is a function of the personal needs of individuals who are doing the purchasing.[55] Specifically, he found that the purchase decision outcome is related to the personal need for professional recognition and career advancement and the social need to satisfy those peers and business associates in the firm who are potential users of the purchased good. Similarly, Bonoma and Zaltman conclude that empirical psychological research on motivation suggests the principle of individual self-interest is one of the most important variables in industrial purchasing behavior.[56]

Risk is more important in industrial purchasing than consumer purchasing because, first, in consumer purchasing there is no discrepancy between macro and micro levels of risk as is the case in industrial purchasing (firm risk versus individual risk) and, second, because *individual-level* industrial purchasing risk tends to be greater than consumer purchasing risk.[57] Consumer purchasing entails lower perceived risk because the undesirable consequences of a bad consumer purchase tend to be less than the undesirable consequences of a bad industrial purchase *at the individual level.* For instance, even with a consumer purchase that is large relative to family income, such as a new car purchase, the consequences of a bad decision will be a temporary reduction in yearly income. Similarly, the undesirable consequences of an industrial purchase of a mainframe computer *to the firm* are also adequately summarized as a loss of income or, more specifically, lost profits. However, the consequences of an industrial purchase of a mainframe computer *at the individual level* can be much more severe. In particular, the risk of a bad purchase at the individual level may entail loss of professional reputation outside the firm; loss of social stature and respect among business peers within the firm, especially if they are users of the product; loss of intrafirm reputation among management superiors, thus jeopardizing career prospects; and perhaps complete loss of employment.

Company image and reputation can reduce perceived risk at both the firm and individual levels. For example, as Chapter 4 will show, it is not simply that the name of IBM means greater reliability and dependability at a company level but that the name of IBM represents a safe, risk-free purchase at an individual level. A data processing manager is not likely to be fired for picking an IBM product that does not work, but the purchase of a product that fails that was produced by an unknown manufacturer can jeopardize much in the way of individual-level career goals.

In addition to the importance of the reduction of risk to all customers as a general product-augmentation strategy, risk can also serve as a basis for customer segmentation. Because the risk an individual associates with a purchase is dependent on his or her perception, individuals can be distinguished according to risk. For instance Levitt found that perceived purchase risk varied with the functional area of the buying-center member.[58] Bonoma and Shapiro concluded that customer attitude and response to risk (a buyer's "risk-management strategy") as well as the risk of the purchasing situation are important segmentation variables.[59] Wilson found that customers could be categorized along a continuum between "risk-neutral" buyers exhibiting "normative" purchasing behavior, and risk-averse buyers "avoiding uncertainty" and exhibiting "conservative" purchasing behavior.[60] Finally, Wilson, Mathews, and Sweeny identified four groups of buyers according to two dimensions of risk-reduction patterns: the source from which information is gathered to reduce risk (external versus internal) and the type of risk buyers attempted to reduce (uncertainty: the probability of failure; or consequences: the outcome of failure).[61]

Risk-based customer segmentation in any of these forms can be an important component or even the central focus in a product segmentation strategy in that it allows the seller to shape the product benefits it offers to more precisely fit the risk profile of the customer segment. Indeed, Chapter 6 shows how Control Data Corporation better matched its product to its target segment as a result of better understanding the unusually low levels of purchase risk perceived by its customer segment. Chapter 5 provides an example of how differences in perceived risk, due to differences in customer self-sufficiency, provide a basis for customer segmentation that, in turn, facilitates price discrimination. In the final analysis, because information plays an elevated role in a two-dimensional heterogeneous markets, risk that derives from inadequate information becomes an important focal point of strategic market behavior.

CHAPTER SUMMARY

Using the idea of two-dimensional market heterogeneity, this chapter brought together in one framework the individual business strategy concepts of market behavior developed in Chapter 2. This chapter then identified and developed the implications of two-dimensional market heterogeneity for the functioning of markets. In particular, it showed that two-dimensional market heterogeneity leads to a more complex view of the function of markets to include their function of matching products with customers, through the superordinate mechanism of market information toward the goal of not simply efficiency but also effectiveness.

The ideas developed in the first part of this study can be summarized in the proposition that the essence of market strategy lies in the ability to take advantage of two-dimensional market heterogeneity and its implication for information. In particular, the fundamental sources of demand-based strategic advantage rest in the capacity to exploit product heterogeneity, customer heterogeneity, and information inadequacy. These in turn imply a variety of more specific behavior forms that are often interrelated. For instance, offering superior customer service exploits potential product heterogeneity in the form of general-market product augmentation and exploits information inadequacy in the form of customer risk reduction; product bundling exploits both customer heterogeneity and product heterogeneity in the form of price segmentation and product augmentation. The next three chapters show that these ideas are empirically grounded in actual market behavior and that, in particular, they form a useful heuristic for identifying the strategic mechanisms underlying observable patterns in the market for mainframe computer products.

NOTES

1. W. Alderson, *Dynamic Marketing Behavior: A Functionalist Theory of Marketing* (Homewood, IL, Irwin, 1965), p. 31.

2. It is the address models of differentiation, and in particular the "quality" or "vertical differentiation" type (that paramatize income), that point to the *potential* relatedness of product augmentation and price segmentation.

3. J. M. Clark, *Competition as a Dynamic Process* (Washington, DC, Brookings Institution, 1961); also see J. M. Clark, "Toward a Concept of Workable Competition," *American Economic Review*, 30 (June, 1940).

4. Abbott, *Quality and Competition*.

5. B. P. Shapiro, "Can Marketing and Manufacturing Coexist?," *Harvard Business Review*, 55, No. 5 (September–October 1977), pp. 104–114. Also see W. Kiechel, "Three (or Four, or More) Ways to Win," *Fortune* (19 October 1981), pp. 181, 184, 188.

6. Porter, *Competitive Advantage*, pp. 16–18, 43.

7. T. Levitt, *Marketing for Business Growth*, 2d ed. (New York, McGraw Hill, 1974), p. 36.

8. W. K. Hali, "Survival Strategies in a Hostile Environment," *Harvard Business Review*, 58 (September–October, 1980), pp. 75–85.

9. L. W. Phillips, D. R. Chang, R. D. Buzzell, "Product Quality, Cost Position and Business Performance: A Test of Some Key Hypotheses," *Journal of Marketing*, 47 (Spring, 1983), pp. 26–43.

10. P. W. Farris, D. J. Reibstein, "How Prices, Ad Expenditures and Profits are Linked," *Harvard Business Review*, 57 (November–December, 1979), pp. 173–184.

11. J. Reddy, "Incorporating Quality in Competitive Strategies," *Sloan Management Review*, 21 (Spring, 1980), pp. 53–60. Also see W. C. Smith, "Finding New Opportunities for Profitability in Manufacturing Cost," *Management Review*, 69 (March, 1980), pp. 60–62.

12. C. H. Fine, "Quality Control and Learning in Productive Systems" (Stanford Business School Ph.D. thesis, 1983).

13. S. C. Wheelwright, "Japan—Where Operations Really are Strategic," *Harvard Business Review*, 59 (July–August, 1981), pp. 67–74.

14. Fine, "Quality Control and Learning," Chapter 1.

15. R. D. Buzzell, F. D. Wiersema, "Successful Share Building Strategies," *Harvard Business Review*, 59 (January–February, 1981), pp. 135–144.

16. M. T. Flaherty, "Market Share, Technology Leadership and Competition in International Semiconductor Markets" (Harvard Business School Working Paper, 1982).

17. J. R. Wensley, C. Robin, "Strategic Marketing: Betas, Boxes or Basics?" *Journal of Marketing*, 45 (Summer, 1981), pp. 173–182.

18. The PIMS (Profit Impact of Market Strategy) Program data base is a proprietary research data base updated annually and compiled by The Strategic Planning Institute (SPI), Cambridge, MA.

19. Phillips, Chang, Buzzell, "Product Quality, Cost Position and Business Performance," p. 42.

20. Demsetz, *Economic, Legal and Political Dimensions of Competition*, pp. 5, 7, 11.

21. D. M. Gardner, "The Role of Price in Consumer Choice," in *Selected Aspects of Consumer Behavior: A Summary from the Perspective of Different Disciplines*, prepared for the National Science Foundation (Washington, DC, U.S. Government Printing Office, 1978), p. 415.

22. Alderson, *Marketing Behavior and Executive Action*, p. 101.

23. Dickson, Ginter, "Market Segmentation, Product Differentiation, and Marketing Strategy," p. 4.

24. The marketing multiattribute approach differs from the traditional economic demand function, in which demand is a function of price given a set of fixed product attributes, in that it incorporates these "givens" as variables within the demand function. The demand function associated with the multiattribute approach is similar to the hedonic model of demand underlying hedonic pricing analysis in

economic theory except that price is not separated from the other product attributes but included with the vector of attributes.

25. See, for instance, Lancaster, *Variety, Equity and Efficiency.*

26. Note that in the hypothetical case of perfectly *evenly* distributed preference points a product movement in *any* direction would increase sales.

27. Note that the strategic import of information inadequacy also lies in the ability to contribute to its cause through, for instance, frequent product changes that, as noted in the next chapter, tend to exacerbate inadequate information conditions.

28. Although Scitovsky identified the important role of information in customer price sensitivity a decade prior to Stigler's seminal "Economics of Information," it was not until this later work that the implications of imperfect information had a substantial effect on economic thinking; see T. Scitovsky "Ignorance as the Source of Oligopoly Power" *American Economic Review* (May, 1950), pp. 48–53, and G. Stigler, "The Economics of Information" *Journal of Political Economy,* No. 3 (June, 1961), pp. 213–225.

29. This term refers to imbalanced information levels between the buyer and seller; see G. Akerlof, "The Market for Lemons: Quality, Uncertainty and the Market Mechanism," *Quarterly Journal of Economics* 84, No. 3 (August, 1970), pp. 488–500.

30. T. Nagle, "Economic Foundations for Pricing," *The Journal of Business,* 57, No. 1, Pt. 2 (January, 1984), p. 4.

31. B. Klein, K. B. Leffler, "The Role of Market Forces in Assuring Contractual Performance," *Journal of Political Economy* 89, No. 4 (August, 1981), p. 624.

32. Nagle, "Economic Foundations for Pricing," p. 8.

33. C. Shapiro, "Premiums for High Quality Products as Returns to Reputations" *The Quarterly Journal of Economics* (November, 1983), p. 660.

34. B. T. Ratchford, G. T. Ford, "A Study of Prices and Market Shares in the Computer Mainframe Industry," *Journal of Business,* 49 (1976), pp. 125–134.

35. C. Wilson, "The Nature of Equilibrium in Markets with Adverse Selection" *Bell Journal of Economics,* 11 (Spring, 1980), pp. 108–130.

36. P. Nelson, "Advertising as Information Once More," in *Issues in Advertising,* edited by D. Tverck (Washington, DC, American Enterprise Institute, 1978).

37. The first work along these lines focused on the product rather than product attributes; see P. Nelson, "Information and Consumer Behavior," *Journal of Political Economy,* 78 (March–April, 1970), pp. 311–329. For reformulations based on attributes see P. Nelson, "Advertising as Information Once More," and "Comments" on "The Economics of Consumer Information Acquisition," *Journal of Business,* 53, No. 3 (July, 1980), pp. 163–165, as well as L. Wilde, "The Economics of Consumer Information Acquisition," *Journal of Business,* 53, No. 3 (July, 1980), pp. 143–158.

38. M. R. Darby, E. Karni, "Free Competition and the Optimal Amount of Fraud" *Journal of Law and Economics,* 16 (April, 1973), pp. 67–88.

39. Nagle, "Economic Foundations for Pricing," p. 51.

40. J. S. Bruner, "On Perceptual Readiness," *Psychological Review,* 64 (March, 1957), pp. 123–52.

41. For the important role of perception as the prime construct in understanding the consumer's reaction to price see the early work of Oxenfeldt: A. R. Oxenfeldt, "Product Line Pricing" *Harvard Business Review,* 44 (July–August, 1966), pp. 137–44; "How Housewives Form Price Impressions" *Journal of Advertising Research,* 8 (September, 1968), pp. 9–17; A. R. Oxenfeldt, D. Miller, A. Shuchman, C. Winick, *Insights into Pricing: From Operations Research and Behavioral Science* (Belmont, CA, Wadsworth, 1961).

42. T. Scitovsky, "Some Consequences of the Habit of Judging Quality by Price," *Review of Economic Studies,* 12, No. 31 (1944–45), pp. 100–105.

43. Gardner, "The Role of Price in Consumer Choice," pp. 424–426.

44. K. B. Monroe, "Buyers' Subjective Perceptions of Price" *Journal of Marketing Research,* 10 (February, 1973), p. 72.

45. Ibid.

46. Porter, *Competitive Advantage,* pp. 143–145, 139, 154–155.

47. Shapiro, "Premiums for High Quality Products as Returns to Reputations," p. 659.

48. Ibid., pp. 659, 660.

49. Ibid., p. 678.

50. D. F. Cox, *Risk Taking and Information Handling in Consumer Behavior* (Boston, Harvard Business School, 1967). See also J. W. Taylor, "The Role of Risk in Consumer Behavior," *Journal of Marketing,* 38 (April, 1974), pp. 54–60.

51. R. M. Hill, R. S. Alexander, J. S. Cross, *Industrial Marketing,* 45th ed. (Homewood, IL, Richard D. Irwin, 1975).

52. R. M. Cardoza, "Segmenting the Industrial Market," in *Marketing and the New Science of Planning—Fall Conference Proceedings,* edited by R. L. King (Chicago, American Marketing Association, 1968), pp. 433–440.

53. Bonoma, Shapiro, *Segmenting the Industrial Market,* p. 83.

54. W. C. Howard, *Selling Industrial Products* (New York, Norton, 1973), p. 283.

55. F. W. Webster, Jr., "Interpersonal Communication Salesman Effectiveness" *Journal of Marketing,* 32 (July, 1968), pp. 7–13.

56. T. V. Bonoma, G. Zaltman, *Psychology for Management* (Boston, Harvard Business School, 1981).

57. W. C. Howard, "How Industry Buys," available form Harvard Business School Case Services, order number 3-574-083 (1973).

58. T. Levitt, "Communications and Industrial Selling," *Journal of Marketing,* 31 (1967), pp. 15–21.

59. Bonoma, Shapiro, *Segmenting the Industrial Market,* p. 88.

60. D. T. Wilson, "Industrial Buyers' Decision-Making Styles," *Journal of Marketing Research,* 8 (November, 1971), pp. 433–436.

61. See D. T. Wilson, H. L. Mathews, J. F. Monoky, Jr., "Information Source Preference by Industrial Buyers as a Function of the Buying Situation" (Pennsylvania State University College of Business Administration Working Paper 27, 1975).

II

Two-Dimensional Heterogeneous Market Behavior in the Mainframe Computer Business

4

Product Augmentation and Customer Risk Reduction by IBM in the Market for Mainframe Computer Products

> Success and leadership have gone to IBM largely because of its better total package of customer-satisfying and therefore customer-getting values. The most important part of this package turns out to be something quite profoundly different from the computer hardware itself. It is the . . . application aids, programming services, information-systems advice, and training programs for their customers . . . with which the generic product is so effectively surrounded.
>
> THEODORE LEVITT, originator of
> the augmented product concept[1]

IBM's customer effectiveness and consequent advantageous market posture has been attributed to various sources, from superior hardware performance and thus price/performance competitiveness, to the abuse of market power. This chapter focuses on the applicability of two related sources of market advantage identified in the theoretical discussion of the first part of this thesis: product heterogeneity and information inadequacy. In particular, this chapter shows that an important source of IBM's customer appeal lies in its product augmenting behavior that exploits the effects of information inadequacy and that specifically takes advantage of low buyer sophistication, signals information to the customers, and most importantly, reduces perceived customer purchase risk. The chapter begins with a brief conceptual illustration of how the special context of IBM's behavior logically inclines it away from the use of price-competitive behavior and toward nonprice behavior. It then accumulates evidence, culminating in a new interpretation of a detailed statistical data base on purchasing behavior, supporting the idea that risk-reducing product augmentation is an important source of IBM's market effectiveness.

THE CONTEXT OF MAINFRAME MARKET BEHAVIOR

The Competitive and Customer Context: Behavioral Implications for IBM

As reviewed in Chapter 2, a key strategic motivation for pursuing product augmentation lies in the potential for generating a discontinuity in the demand curve, thereby insulating a firm from the profit-reducing tendency of price competition. In IBM's case, however, several situation-specific variables, such as the structure of its market, the behavior of its customers, and the nature of the threat posed by competitors, makes the need to avoid price competition more intense.

IBM's large market share, relative to the small amount of share threatened by the price-cutting behavior of a new entrant or a small competitor, creates a disincentive for IBM to engage in, or respond to, competitive price cuts by lowering its prices. This is so because IBM typically stands to lose more through self-inflicted losses due to lower margins than it gains in market share threats averted. The greater the disparity between IBM's share and the maximum share a new entrant can capture, the more acute the need to avoid a price-competitive response. Consider Example A in Figure 4.1. A low-priced competitor threatens up to 10 percent of IBM's sales for the next year. To eliminate this threat and protect all its share, IBM could cut its price to a level close to the competitor's price. However, the gain in profit as a result of deterring entry is far less than the loss to profit due to lower contribution margins.

The disincentive to take competitive price actions is intensified by the presence of existing effective augmentation. The same augmentation behavior that protects IBM from competition and allows it to command price premiums also makes price cutting more costly for IBM. In particular, the same band of price insensitivity that insulates IBM from competitors' price cuts also increases the gross losses to IBM if IBM decided to match these cuts. The matching of such price reductions by IBM will result in some level of lost revenue for every customer who would have remained loyal to IBM despite the lower-priced competition. To the extent that this anticipated lost revenue exceeds the expected value of limiting competitive entry and maintaining or increasing market share, a price reduction response is dysfunctional. Hence, a price reduction by IBM from a price level 50 percent over competition, to a level 20 percent over competition, may do little else than reduce IBM's margins without attracting the price-sensitive customers because of the smaller premium still present (as suggested in Example B of Figure 4.1).

The presence of effective augmentation thus increases the potential for self-inflicted losses and widens the disparity between self-induced losses and the gain from limiting a rival's share. Stephen Ippolito's explanation of how he and other small rivals exploit this crack in IBM's strategic armor is an accurate reflection of the fundamental market dilemma of the dominant firm in an otherwise decentralized, unconcentrated industry:

The whole key to this industry is the price umbrella phenomenon. . . . For IBM to cut their price down to the point [where other] vendors will not be at a price disadvantage—to go after the other 10% or 15% [of the customers] that are not willing to pay a premium for the IBM name—they would actually be losing money on the other 85% of the market that are willing to pay a premium for the IBM name.[2]

The fact that existing effective product augmentation exacerbates the net losses from price-cutting behavior, by widening the gap between self-induced losses and gains from averting market share threats, provides support for the idea of an internal inconsistency in mixing a price-competitive strategy and a product-augmentation strategy, at least in the case of the behavior of a dominant market share firm like IBM.

The urgency of price competition avoidance is still further intensified by the

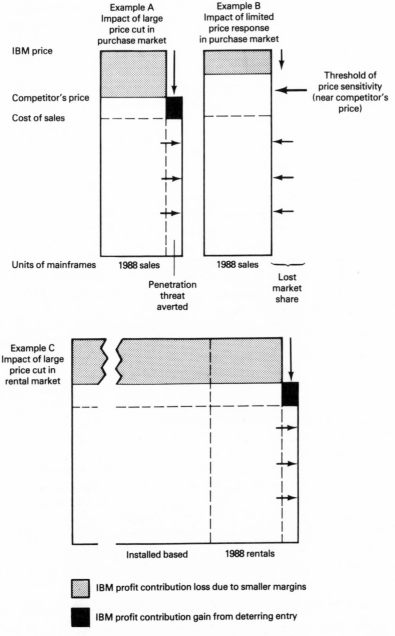

Figure 4.1. IBM's price-cutting disincentive.

presence of a customer rental base (versus a purchase only market). IBM's rental base constrains IBM from cutting prices because the benefit of the price cut, in terms of limited entry or market share protection, may be more than offset by the cost in terms of revenue lost on existing rentals. IBM's capacity to cut price in response to competitive attacks is constrained not only by losses on future rentals lost but by potential losses it would incur on the entire outstanding installed rental base that has accumulated over several product generations. The precise magnitude of these losses is theoretically measurable by summing the revenue lost due to reduced rental rates on every IBM product that customers would have held despite the competitive attack.

Assume only 5% of the total combined value of IBM's installed base plus anticipated 1988 rentals is threatened by low-priced competition (because only the limited number of price-sensitive customers are vulnerable to the lure of lower prices). At a contribution margin of 45%, the maximum loss IBM can suffer is 2.25% of its total product revenue (5% of 45%). If IBM cuts price and reduces margins to 15% to protect share (as illustrated in Example C of Figure 4.1), then the loss is well over half of the contributed profits from the entire installed base plus the 1988 rentals, minus the gain from averting a small market share threat. As Example C suggests, however, the limited gain can hardly justify the self-induced losses from reduced margins. Using the "maximum" criterion of formal decision analysis (the alternative whose worst consequence is best), IBM will choose to not cut price. The larger the rental base, the greater the disparity between profits lost through reducing margins and profits gained by protecting share.

In summary, because of the special competitive context in which IBM is immersed, the urgency to avoid pure price behavior is more acute. In particular, an enormous *relative* market share, a historically large rental base, and the ongoing presence of effective product augmentation, and therefore an already established band of price insensitivity, mean the self-induced losses to IBM of engaging in highly price-competitive behavior can far outweigh any gains from market share threats averted.[3]

Competing through nonprice, product dimensions better allows IBM to secure its share without hurting itself. Product augmentation in particular serves as both an effective defensive strategy and offensive strategy. It is used as a defense against competitive price actions by creating a buffer zone of price levels within which demand remains unresponsive to price differences. By pricing at the top end of this range IBM is able to command a price premium while at the same time forcing price-competitive-orientated rivals into pricing below, not just IBM's price, but, the bottom of the buffer range in order to have any appreciable effect in attracting customers. As an offensive strategy product augmentation engenders customer loyalty and facilitates greater firm stability, which in turn may enhance IBM's basis for image differentiation.

The Product Market Context and Its Behavioral Implications for IBM

The significance of information inadequacy for business policy lies in the diminished effectiveness of price competition and the enhanced importance of nonprice behaviors that exploit this market condition. These points were discussed in the second half of Chapter 3 and need not be repeated here; however, it is important to point out why these business policy implications become all the more significant for dynamic markets, such as the market for mainframe products.

Briefly, the reason the business strategy implications of information inadequacy are more important in dynamic markets is that information inadequacy, its causes, and its effects tend to be more pronounced in dynamic markets. Rapid product change in the computer market leads to inadequacies in market information that otherwise might not occur in less dynamic, more stable industries. Rapid and regular change prevents product markets from reaching maturity levels they might otherwise reach. The duration of the product life cycle is shorter and the cycle rate is faster. Relatedly, the distribution of products according to their life-cycle stage is skewed more heavily toward earlier stages. Products have less time to evolve into commodities and thus tend to be less commodity-like than they would be otherwise.

Similarly, with respect to a given product, the customer base has less time to evolve. In this way the phenomenon of attenuated product life cycles limits the customer's familiarity and experience with products thereby limiting his ability to compare among them. Customer sophistication (the level of competency in understanding, evaluating, and utilizing a product to its greatest potential) is consequently lower than it would be in static markets. Early product stages mean customer behavior tends to be weighted toward the behavior of "early adopters" and thus is less price sensitive than otherwise.

Moreover, rapid product change, especially if coupled with large purchase sizes and relatively infrequent purchase intervals, as in the case of the mainframe industry, suggests a tendency toward first-time purchases. Unlike the first-time purchaser or inexperienced buyer in the stagnant market who can draw on the product information acquired by more experienced buyers to reduce risk, the inexperienced buyer in the dynamic computer market is likely to find less product information from other buyers. Shorter cycles also imply that information has less time to disseminate and is thus more scarce, less available, and perhaps less accurate and reliable. Scarce information implies more costly information, which in turn inhibits the customer's incentive for improving his sophistication.

The technological factor, however, is important not only because of the high rate of change it implies but also because of its implication for the complexity of products as well. Complexity in process technology may slow down the pace at which competitors can imitate innovation, thereby increasing disequilibrium that derives from the rapid rate of product change. Moreover, complexity in product technology inhibits rational price comparison and the ease of processing information useful in purchase decisions. Product complexity thus further exacerbates information inadequacy and contributes to low customer sophistication by not only increasing presence and need for data but also by straining the capacity to process that data. Together the pace of product change and the complexity of products create market conditions inhibiting customers' access to adequate information about products.

The strategic significance to IBM of such market dynamism and resultant information inadequacy lies in the capacity to contribute to this condition through product actions or customer management techniques that interfere with the evolutionary process of markets from dynamic to static, of products from novelties to commodities, and of customers from ignorant to sophisticated. The significance of such intensified conditions of information inadequacy also lies in the ability to exploit its effects (the customer's desire for product information and his perception of purchase risk) through behavior that signals information and reduces perceived risk. The presence and use of such behavior by IBM is the subject of the rest of this chapter.

ACCOUNTING FOR PRICE PREMIUMS

Price competition is not a topic of this study, but this fact is not meant to imply the denial of the important role of price competitive behavior in the mainframe computer industry. Plug-compatible manufacturers (PCMs) both in the mainframe peripherals market and the central processing unit (CPU) market use price behavior to capture and sustain market share.[4] Japanese computer manufacturers use price behavior to exploit cost advantages in gaining entry to mainframe markets. An analysis of the price competitive behavior and strategy of Japanese computer manufacturers is provided by Davidson, who shows that these "firms have shown almost a universal willingness to accept margins and profits lower than those prevailing in world markets."[5]

The importance of competitive pricing for this study lies in the interpretation of price differences among competitors. IBM mainframe products have traditionally been sold at a premium price relative to competitors' products. Sobel explains that to compete against IBM "the industry rule of thumb is that you have a chance if your machine performs ten percent faster and is ten percent lower in price."[6] In the IBM 360 Series mainframe market, DeLamarter observes that, "on the average, IBM's prices were higher." He further observes that, "when confronting IBM products, competitors priced their systems between .10 and 15 percent under the 360's 'price umbrella.' "[7] Testimony by executives of firms competitive with IBM reveal that, historically, competitors have had to price from 5 to 20 percent below IBM's prices.[8]

Ratchford and Ford have conducted empirical research investigating the source of price premiums in the heyday of IBM's technological sophistication, the 360 period. In particular, these authors attempted to determine if premiums were present on IBM mainframes even after "product differences" between the products of IBM and its competitors were removed. More specifically, the authors used hedonic price methodology (a methodology employing the multivariate statistical techniques of factor analysis and regression analysis) to construct hedonic price indexes for differentiated mainframe computer systems.

This method assumes that "observed prices of differentiated products" can be reduced to a "set of implicit prices"[9] or "hedonic" product attribute prices that, in effect, reveal the payments customers would make if the products were not different.[10] Actual products are reduced to equivalent products by expressing them in terms of some common performance measure arrived at through factor analysis that theoretically summarizes the products. In this way Ratchford and Ford, in effect, compared prices of equivalent products and identified price premiums that, in theory, did not reflect product differences.

Because this work was approached from the backdrop of antitrust policy, it attempted to test for the presence or absence of price-competitive behavior. If the presence of substantial price differences between "equivalent products" was found, it was said to indicate the lack of price-competitive behavior. In fact, Ratchford and Ford found that "IBM machines are priced substantially above competing machines of equal performance," and therefore tentatively concluded that IBM exhibited "monopoly power." Moreover, their finding that prices far exceeded marginal cost led them to suggest that "the entire market for general purpose computers is not very competitive."[11]

There are several reasons for arguing for the presence of monopoly power and a lack of competitiveness in the computer industry. The phenomenon of "customer lock-in," for instance, is a compelling reason. But the observed presence of price premiums is inadequate as a reason unless it can be shown that the products being compared have been stripped of *all* differences, not just performance differences. Ironically, therefore, research that concludes IBM is a monopolist because of the presence of price premiums after performance differences are removed also points to an equally valid hypothesis: IBM competes effectively through non-performance-based product augmentation. This explanation is presented here not as a mutually exclusive alternative hypothesis, but as a very important additional factor that any comprehensive and balanced explanation of price premiums on IBM products should recognize.

The answer to the question of monopoly power and lack of competitiveness is not so much of importance in this thesis as Ratchford and Ford's suggestion that "since a substantial percentage of users are still willing to employ IBM machines even though their relative price is very high, the implication is that IBM offers customers something to induce them to pay a substantial premium for an IBM machine."[12] Based on their finding of substantial price premiums even after product differences are removed, Ratchford and Ford offer this explanation:

Apparently IBM offers substantially more (or higher quality) non-hardware services than do its competitors. Possible examples of these non-hardware services are superior after-sales service, more product reliability, better software, and more assistance in developing applications of the machine. The fact that the large price differential between IBM and its competitors persisted after the period of this study . . . suggests that other manufacturers have not been able to make their non-hardware offerings as attractive as IBM's. Apparently IBM has maintained a long-standing comparative advantage in the area of non-hardware services.[13]

Ironically such an observation points to the ineffectiveness of their work in achieving one of its basic aims: the removal of all "product" differences. The problem lies in their limited view of the product and, in particular, the implicit concept of a product only as the "core product," the "thing itself," rather than the total bundle of benefits it provides to customers, as the augmented product concept of marketing literature requires. The limited definition of a product in terms of physical characteristics was perhaps a necessary practicality for this type of research in light of the immense difficulties in reducing the core characteristics of mainframe systems to a common performance measure, nonetheless attempting to account for other, nonphysical, differences. The "nonhardware services" that they invoke as an explanation of price premiums even after "product differences" are removed are, in the view of this author, a critical part of those products differences.

Moreover, even within the limited product definition used by Ratchford and Ford, there is serious doubt as to the effectiveness of their work in eliminating product performance differences across different manufacturers. Industrial economist and prominent computer industry expert G. W. Brock, in a comment to Ratchford's and Ford's article, is highly critical of the premium they found (between 40 and 50 percent) although the presence of differentiation via nonhardware services is not questioned. Brock concludes:

There is serious doubt as to whether even a very useful study of the prices of various computers as a function of only two characteristics can be accurate enough to distinguish between manufacturers. The additional problem of inaccurate and inconsistent data makes it highly doubtful that any confidence can be placed in the [Ratchford and Ford] results.[14]

Ratchford and Ford respond to Brock's comment by acknowledging the data errors uncovered by Brock but arguing that their "earlier conclusions were not affected in any essential way by the error."[15] Michaels, using an alternative sample and a modified approach to the problem, reaches different conclusions about price premiums. Using hedonic price equations Michaels, finds that "brand premiums on IBM equipment are generally smaller"[16] than those found by Ratchford and Ford. While both Michaels and Ratchford and Ford consider the role of tangible nonhardware services, neither consider the possible role of inadequate market information as a source of price differences among the equivalent products. Also, more specifically, none of the authors consider the economic arguments that the premiums might also reflect the opportunity cost of exploiting consumer ignorance or the fair returns to reputation (as discussed in Chapter 3).

The debate concerning the size of IBM's price premium notwithstanding, one certain contribution of all of the above noted mainframe industry research is the not surprising finding that it is extremely difficult to compare prices of heterogeneous computer "products," even if the definition of product is limited to the core product. This difficulty, moreover, refers to the capacity of scholarly research after the fact of purchase has taken place. Once the more limited capacity of the actual customer before an actual purchase is considered, and a broader definition of the product is adopted, the difficulty of reducing a purchase decision to a price comparison is more fully appreciated.

A second contribution of both the Michaels and Ratchford and Ford studies is the identification of the presence of price premiums on IBM mainframes (despite the lack of scholarly opinion as to the approximate size of that premium). The debate as to premium size should not detract from the more basic conclusion that premiums are present and the even more certain conclusion that IBM does not offer *lower* prices on equivalent performing machines. Superior hardware performance therefore cannot be easily envoked as an adequate explanation for IBM's market success.

In fact, this conclusion is not inconsistent with IBM's own findings concerning the price/performance competitiveness of its products. Court trial disclosures in both the Telex case and the most recent Justice Department case brought against IBM reveal the results of price/performance comparisons between IBM products and competitive products by the one agent least likely to underestimate the performance of IBM products: IBM itself. Levitt observes that "subpoenaed documents in the IBM–Telex litigations disclose IBM judging any number of its competitors' computer products better than its own." He concludes that "it has become perfectly clear that (IBM) is not necessarily the one whose computers are always the best or the cheapest."[17] More recent, and more convincing, is the information made available in the Justice Department case. In particular, "IBM-confidential" documents disclosed in the trial record of the *U.S. v. IBM* case reveal that IBM itself judged what was at the time held to be the premiere mainframe series, the IBM 360, as, at best, merely equal to competitive offerings and, for a large percentage of the 360 models, clearly inferior to competitive equivalents.

During the period of the 360 product generation IBM conducted sophisticated evaluations of its mainframes relative to competitors' mainframes. The evaluation criterion was price/performance. The results of these evaluations were reported to the Data Processing Group (the marketing function in IBM) in "Quarterly Product Line Assessments" (QPLAs). These evaluations are summarized in Table 4.1, which provide IBM's assessments of each of its commercial 360 models at year-end across four years. In over half the cases (58%) IBM rates its 360 system only equal to the price/performance competitive offerings. In over a third of the cases (38%) IBM rates its system as offering inferior price/performance. In only a single case (4%) IBM judges its mainframe better than competition.

Furthermore, the lack of price/performance superiority as well as the presence of an advantage in nonperformance product attributes (such as service, support, and reliability) seem to be confirmed by customer perceptions of IBM's mainframes relative to competitive mainframes of this period. Figure 4.2 shows that, in a survey of 389 mainframe customers, IBM was ranked first by at least 70 percent of mainframe users in all nonperformance categories. In price/performance, however, the results are reversed; non-IBM computer manufacturers were given superior ratings by over two-thirds of all customers.

These "nonhardware services" benefits are the subject of the remaining two sections of this chapter. The next section discusses the role of IBM's marketing policy of product bundling as a vehicle to augmentation along nonhardware dimen-

Table 4.1 IBM's Internal Product Assessments (IBM 360 Systems relative to competitive offerings)

Model	1967	1968	1969	1970
360/20	−	−	−	−
360/25	0	0	0	−
360/30	0	0	0	−
360/40	0	0	0	−
360/50	+	0	−	−
360/65	0	0	0	0

Source: U.S. v. IBM. These data are compiled from fourteen plaintiff exhibits of various dates. Identical or similar data can be found in an "IBM Marketing Report," Plaintiff Exhibit 3360-B. These data are also available in Appendix C, Table 45, p. 346 in *Big Blue* by De Lamarter.

Note: + means superior
     0 means equal
     − means deficient

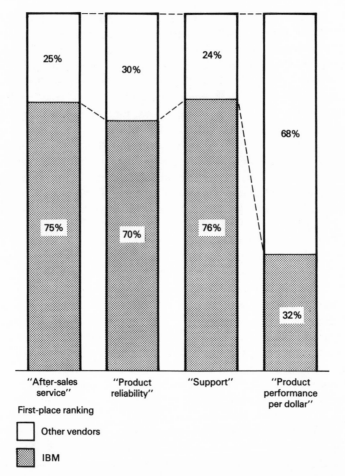

"After-sales   "Product   "Support"   "Product
service"   reliability"   performance
                                                          per dollar"
First-place ranking

☐ Other vendors

▨ IBM

Figure 4.2. Customer survey ratings (for the IBM 360 product generation and 360 competitors). *Source:* R. McLaughlin, "Monopoly is Not a Game," *Datamation* 19 (September, 1983).

sions. The last section presents statistical data on buying behavior indicating the important role of this form of augmentation in IBM's strategic effectiveness.

RISK REDUCTION THROUGH PRODUCT BUNDLING

In the computer industry, bundling is the offering of two or more hardware, software, or service elements of a data processing system under a single pricing plan, without detailing the pricing of the individual products comprising the bundle. Actually, the wording of IBM sales and lease contracts treated the price customers paid for the total bundled system as the hardware price; all the other elements of the bundle were considered to be given away free. The importance of this marketing practice lies in its utility for exploiting both product dimensionality (and thus facilitating product augmentation) and customer dimensionality (facilitating price segmentation). This chapter is concerned with the role of bundling in product-dimensional strategy only.

All product augmentation is, in a sense, bundling, for the essence of the augmented-product concept is to view the product as more than merely the core physical product but as the full bundle of benefits that surround the core. From the augmented-product perspective, "a computer is not simply a machine for data storage, processing, calculation or retrieval,"[18] but "a complex cluster of value satisfactions that include education, training, hands-on help, continuing advice and quick availability for emergency situation."[19] More generally, IBM's mainframe product bundle consists not only of hardware but of "an unspecified amount of support and software."[20] "Support" can include installation service, conversion assistance, maintenance, systems design, systems engineering, the identification of application needs, applications development, actual software programming, as well as customer education and training in any of these areas.

Support is not only bundled in the legal sense that prices are unspecified but the support itself is unspecified. This approach is designed to contribute to an image of IBM as committed to making sure its system satisfies customer desires and meets expectations no matter what this might entail. As IBM executive Faw testifies, IBM's product

offering was much more than mere provision of a machine and its maintenance. Then as now, it included methods and analysis of existing administrative procedures, conversion, education, systems service and a continuing stream of advice and counsel for extensions and an implied guarantee the system would work as promised.[21]

Similarly Levitt, in one of his few references to the computer industry, presents IBM's behavior as a paradigm case of product augmentation:

It [IBM] bundled the software right into the product offering at a single set price, so that the customer was assured that the equipment would indeed be programmed to do the promised job. It designated installation facilities for the customer, redesigned his entire data collection and reporting systems, trained his data processing people, took the shakedown cruise, and then later developed new EDP applications to help the client even more.[22]

Offering all the hardware, software, and service as a total system at one price eased customers' budgeting difficulties associated with an expensive and complex product. Moreover, the nature of the service components in the bundle—an open-ended commitment to customer support—reduced customer concern over unforeseeable problems. The testimony of IBM executive Welke reveals both these elements of IBM's bundling strategy:

On the one hand, it gave the users a predictable cost that they could budget against. They knew that their system would cost them "X" number of dollars a year or per month, and they could budget that amount and predict it. And by the same token, they also knew that the *undefined problems* that existed in data processing, in their computing world, would be covered as well. . . . If I know that education, maintenance, the various support services are mine for the asking . . . that in whatever quantity I might need them they will be made available, then I have a predictable cost that I can allocate to computing. I can say that . . . my installation, my computer is going to cost me . . . $15,000 a month . . . and all of these things are included. It will be an operating system. It will do my job for me. It is the solution to that data processing problem.[23]

IBM's "total solution" marketing approach and bundling practice exploit the contextual product and customer conditions in which it is immersed. As discussed earlier in this chapter, the complexity and rapidity of change in product technology diminish the customer's sophistication in evaluating, selecting, and using mainframe products. Bundling, by reducing a complex range of hardware, software, and support into one product package with a single price, simplifies the purchasing decision and thus appeals to the lack of buyer sophistication. It is not that bundling and total-solution marketing improve customer sophistication but that it assuages customer insecurity in making a purchase[24] and encourages the customer to accept information signals such as reputation in the place of specific product information. As Brock points out, "the practice of bundling hardware prices with free software and support services encourages the customer to evaluate total manufacturing capabilities rather than individual products."[25] In short, IBM sells "IBM," not just computers.

DeBruicker and Summe argue that IBM's appeal to lack of customer sophistication is the critical factor in its market success:

A look at the history of the mainframe computer industry's environment tells us much about inexperienced customers and strategies that serve them. . . . IBM dominated this market from the outset because it had tailored its marketing strategy to the inexperienced buyer. Its marketing program included complete systems of reliable (rather than technologically advanced) hardware and software; a wide product line permitting future upgrading; extensive human resources for installation, education, service, and account management; rental options with liberal system-upgrading privileges and fees quoted on a whole-system basis. . . . IBM's was a formidable strategy, not so much because of the enormous resources behind it but because it was tailored to a market of inexperienced generalists.[26]

An important element of IBM's strategy that helps boost a customer's confidence and reduce his concerns in purchasing an IBM mainframe is the unique composition of IBM's total field force and, in particular, the large role of non–sales field personnel. The field force consists of the IBM "representatives" who deal directly with customers. While it is true that IBM has a large number of "marketing representatives" (salespeople) in relative terms (and this no doubt contributes greatly to its customer effectiveness), what is not conventionally recognized is that the personnel concerned directly with sales constitute only approximately one-third of its total field force. For instance, even in the period when IBM's typically dominant sales function was most prominent, the 360 period, over 60 percent of IBM's field personnel were not sales personnel (Figure 4.3). Further, over half of the total field force were dedicated to systems installation. And a substantial number were dedicated to "field engineering," a field function that did not even exist at other computer manufacturers.

The "field engineer"[27] is technically a customer-support position as opposed to a sales position, but the role of the field engineer is inextricably intertwined with creating and sustaining sales. He is viewed as "part of the marketing team." He provides the technical expertise that allows salesmen to offer workable solutions

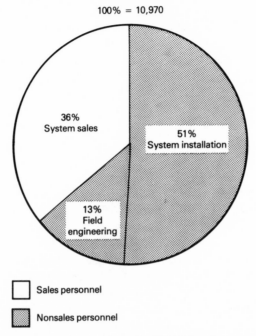

Figure 4.3. Composition of IBM's field force (for IBM 360 series systems). *Source: U.S. v. IBM,* Defense Exhibit 52.

to customer problems and actually helps prepare purchase proposals offered customers.[28]

The field engineer works with both existing customers before the sale and potential customers after the sale to define their data processing needs and requirements, design the system, customize generic IBM software to fit those needs, develop approaches to application problems, actually help write specific application programs, and provide formal and informal training in both hardware use and programming.[29] He has "the implied responsibility of . . . developing systems to make sure that the machine was put to good use"[30] and to make "sure that the customer was indeed implementing the targeted applications" and "that the machine was . . . performing properly."[31] In short, he was the key link in the IBM field force who assured that products sold to the customer actually did solve his problems. The effect of the conspicuous presence of IBM's field engineering personnel on the customer's perception was the creation and reinforcement of an image of IBM as willing to provide help of whatever type and in whatever amount necessary to solve the customer's problems.

The importance of the "unspecified support" component in IBM's bundled offering does not so much lie in the notion that IBM actually did provide extensive support but in the customer perception that IBM was willing to provide support in "whatever quantities" needed. DeLamarter argues that "the amount of assistance that IBM was capable of delivering . . . at no extra charge . . . was essentially limitless."[32] It is, however, not the fact of "limitless" support, but the belief by the customer that this is the case that is important. In other words, it is not the accuracy of the customer's perception, nor the reliability of the information signaled by IBM's reputation, but the perception itself, the reputation itself, the signaling phenomenon itself that are important to IBM's marketing effectiveness. It is, in the end, the customer's belief in IBM's own marketing motto "we sell solutions, not computers" that sells computers. It is these customer perceptions underlying mainframe product purchasing decisions that form the raw material of the data base reviewed in the next, final section.

STATISTICAL EVIDENCE ON MAINFRAME PRODUCT PURCHASING BEHAVIOR

This section uses a rich and detailed statistical data base[33] that is unusual for industrial market research in general and unique to computer mainframe market research in specific. This data base contains the results of an extensive telephone and questionnaire survey conducted by twenty-five full-time research analysts at the management consulting firm Booz Allen and Hamilton under the Supervision of Rowland Morierity, with the cooperation of the Harvard Business School, American Telephone and Telegraph (AT&T), and the Marketing Science Institute. In the rawest form the data consist of completed, detailed questionnaire and telephone inquiry logs that record the responses of 300 customers or purchasing "Decision Making Units" (DMUs), selected through stratified random sampling from Dun & Bradstreet companies covering five industry sectors and three size classifications. This large quantity of detailed raw data is organized into a data base containing several types of data on purchasing behavior in the market for mainframe products.

Although the Morierity Data Base was compiled for, and is focused on, providing "market segmentation" information, it contains three types of data that are here shown to be useful in understanding the role of a general-market, product-augmentation strategy that exploits the effects of information inadequacy. In particular, although the data base's primary focus is the provision of information organized by ten market segments, its utility to this study is its provision of information on the market in general and on the market divided by IBM customers and non-IBM customers.

Data Used

The first of the three types of data useful to this thesis are customer saliency ratings of buying criteria. Purchasing DMUs rated thirty-three different possible criteria for evaluating and selecting a mainframe product. Specifically, on a scale of 1 to 6, they rated how "important" each of thirty-three buying factors were in their purchasing decisions. They also rated, on the same scale, the extent to which they thought each factor "varied" among competitive products within the industry. These two sets of ratings for each factor were then combined to assess how "determinant" or "salient" each factor was in a purchasing decision. More technically, the "determinancy" rating or "saliency" rating is a function of the "importance" rating and the "variability" rating, and specifically, the mathematical product of the two.

Because a high saliency rating depends upon the presence of both high importance and high variability, a factor is not highly determinant, even if it is important, unless it is also variable. In general, for instance, even if price is rated very important as a factor in purchasing, if price does not vary among the products being considered, it is not highly determinant in the buying decision. For example, among the consumers of an airline service, safety is no doubt very important. Because the variability among airline safety is not typically thought to be significant, however, safety is not a highly determinant, or salient, factor in choosing which airline company to use for a particular flight.

Because customers may perceive rating scales differently, the data were normalized. Before the determinancy rating was calculated, each customer's ratings for variability and importance were normalized about a mean of 6 (by subtracting from each customer's rating the mean of all his ratings and adding 6 to that value so the normalized ratings were all positive). After normalization, therefore, the determinancy rating centered about 36.

The second of the three types of data useful to this study are the determinancy ratings of "macroattributes" or, in other words, aggregated buying factors. More precisely, the thirty-three original attributes are aggregated into a set of fourteen macroattributes using multivariate procedures of data correlation and clustering. Of the original thirty-three attributes, twenty-four are combined into nine macroattributes, another five of the initial thirty-three attributes are retained as originally defined, and the remaining four, exhibiting no stability across clustering methods and showing a lack of clarity in respondents' perception after follow-up interviews, are dropped from the data base. The commonsense logic of the statistical procedures used in organizing the original factors into the smaller manageable set of fourteen macroattributes is illustrated in Figure 4.4.

The third type of data useful in this study is actually several specific categories of data, all of which are contained in the broad Morierity Data Base category of "behavioral profile data." Profile data includes customer importance ratings (not saliency ratings) of factors not included in the original thirty-three purchasing factors. Profile data also consists of various statistical facts describing the customers that are expressed in dollars, product units, number of customers, or percentages of these. Unlike most of the other data, profile data is available for IBM and non-IBM customers separately, and herein lies the utility of this type of data for this study.

Most of the remainder of this chapter interprets these three types of data selected from the Morierity Data Base to show the importance of risk-reducing product augmentation in the market for mainframe products in general and to show that such behavior is an important source of IBM's customer appeal in particular. This data not only sheds doubt on the credibility of arguments that attribute IBM's market effectiveness to superior price/performance competitiveness, but it also lends empirical support to the conclusion implied by the phenomenon of an information-poor marketplace (caused by rapid product change and product complexity). In particular, it suggests that the strategic use of the effects of information inadequacy (information signaling, exploiting low buyer sophistication, and reduction of perceived purchase risk) is an important source of market effectiveness.

Original purchase criteria actually rated   Clustered macroattributes

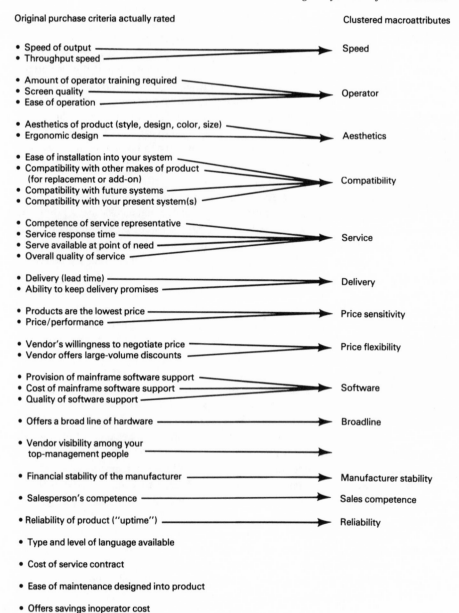

- Speed of output ————————————————————————→ Speed
- Throughput speed ————

- Amount of operator training required ———
- Screen quality ————————————————————→ Operator
- Ease of operation ————

- Aesthetics of product (style, design, color, size) ———
- Ergonomic design ————————————————————→ Aesthetics

- Ease of installation into your system ———
- Compatibility with other makes of product ———
  (for replacement or add-on)   Compatibility
- Compatibility with future systems ———
- Compatibility with your present system(s) ———

- Competence of service representative ———
- Service response time ———
- Serve available at point of need ————————————→ Service
- Overall quality of service ———

- Delivery (lead time) ————————————————————→ Delivery
- Ability to keep delivery promises ————

- Products are the lowest price ————————————————→ Price sensitivity
- Price/performance ————

- Vendor's willingness to negotiate price ————————————→ Price flexibility
- Vendor offers large-volume discounts ————

- Provision of mainframe software support ———
- Cost of mainframe software support ————————————→ Software
- Quality of software support ———

- Offers a broad line of hardware ————————————————→ Broadline

- Vendor visibility among your
  top-management people ————————————————————→

- Financial stability of the manufacturer ————————————→ Manufacturer stability

- Salesperson's competence ————————————————————→ Sales competence

- Reliability of product ("uptime") ————————————————→ Reliability

- Type and level of language available

- Cost of service contract

- Ease of maintenance designed into product

- Offers savings inoperator cost

Figure 4.4. Data aggregation path. *Source:* Moriarty database.

Factors Determining Purchasing Behavior Decisions

Customer Support and Product Reliability
Reviewing the rankings of the determinancy ratings of the original thirty-three purchasing criteria listed in Table 4.2 reveals two patterns. The first pattern concerns the importance of service or support factors in purchasing decisions. Specifically,

the top five rankings, the five most determinant factors in purchasing decisions, are all related to customer service: "service response time," "service availability at point of need," "overall quality of service," "quality of software support," and "competence of service representatives." If the top quintile of rankings is considered, the related dimension of "reliability of product" is added to the list. The entire top quintile of rankings therefore consists exclusively of customer support and product reliability factors.

The same pattern emerges from the clustered data (which, as discussed, is arrived at by aggregating the original thirty-three factors into fourteen macrofactors). Table

Table 4.2 Ranking of Factors Determining Mainframe Product Purchasing Decisions

	Purchasing Criteria Ranking	Determinancy Rating	Purchasing Criteria Description
Most	1	46.46	Service response time
determinant	2	45.57	Service available at point of need
	3	45.46	Overall quality of service
	4	44.49	Quality of software support
	5	44.41	Competence of service rep
	6	43.43	Reliability of product
	7	40.26	Ease of installation into your system
	8	40.23	Compatibility with your present system
	9	39.92	Compatibility with future system
	10	39.92	Financial stability of the manufacturer
	11	38.78	Price/performance
	12	38.74	Provision of mainframe software support
	13	37.75	Ease of maintenance designed into product
	14	37.37	Ability to keep delivery promises
	15	36.92	Delivery (lead time)
	16	35.97	Compatibility with other makes of product (for replacement or add-on)
	17	35.72	Ease of operation
	18	36.65	Throughput speed
	19	35.63	Cost of mainframe software support
	20	35.32	Cost of service contract
	21	35.31	Speed of output
	22	35.30	Salesperson's competence
	23	34.62	Type and level of language available
	24	34.21	Offers a broad line of hardware
	25	32.18	Amount of operator training required
	26	31.49	Screen quality
	27	30.51	Products are the lowest price
	28	29.39	Ergonomic design
	29	29.15	Vendor's willingness to negotiate price
	30	29.07	Offers savings in operator costs
	31	28.42	Vendor visibility among your top-management people
	32	26.37	Vendor offers large-volume discounts
Least determinant	33	25.67	Aesthetics of product (style, design, color, size)

Source: Morierity Data Base.

Table 4.3 Macroattribute Rankings

Rank	Variable	Determinancy Rating		
		Mean	Minimum	Maximum
1	Service	45.47	20.5	72.2
2	Reliability	43.43	18.7	70.7
3	Manufacturer stability	39.92	4.2	66.8
4	Software	39.62	10.6	67.6
5	Compatibility	39.09	13.1	91.0
6	Delivery	37.15	6.6	75.5
7	Speed	35.48	10.6	70.7
8	Absolute price	34.64	13.4	76.1
9	Broad line	34.21	11.6	75.5
10	Operater	33.13	8.9	59.7
11	Sales competence	32.18	8.9	76.6
12	Visibility among top management	28.32	4.5	60.5
13	Price flexibility	27.76	4.2	69.4
14	Aesthetics	27.53	6.0	51.4

Source: Morierity Data Base.

4.3 reveals that customer "service" is the absolute most determinant factor in purchasing behavior, followed by "reliability," the second most salient factor. Hence, again, customer service followed by reliability comprise the top quintile of rankings.

Both sets of data (original and clustered) reflect responses for 300 DMUs over the entire range of buyer companies in the sample—small, medium, and large in size, representing all of the sampled buyer industry sectors. The high absolute rankings of the saliency ratings for support and reliability point to the high sensitivity to these factors among customers in general. This observation confirms the use of product augmentation along service and reliability dimensions as a strategy of broad market appeal and thereby serving as a basis for general market product augmentation.

This observation of high support and reliability ratings also points to the validity of IBM's support-orientated marketing approach in general and bundling policy of offering unspecified support in particular. Moreover, customer support and product reliability and, more specifically, all the factors listed in the top quintile of the original thirty-three factors are product dimensions, the augmentation of which tends to reduce the customer's perception of risk. This points to the importance of risk reduction as a tool of strategy in the market for mainframe products. (More risk-specific data is presented later.)

Seller Stability
By including the third ranked factor in the top quintile of the fourteen macroattributes, "manufacturer stability" is added to the list (Table 4.3). The importance of a seller's stability in purchasing behavior points to the significant role of information signaling in the market for mainframe products. In particular, it signals both specific and general information.

Firm stability specifically indicates the dependability of the vendor as a long-term

supplier. Data processing systems are typically very costly relative to other investments as well as highly integral to the functioning of a customer's business. Moreover, changing to another vendor's system entails substantial costs resulting from
installation, software conversion, retraining, and the disruption to all aspects of the
business that depend upon data processing services. The facts that IBM is one of the
oldest vendors in the market, is the largest vendor in the market, and has a proven
track record of excellent financial performance and consistent steady growth all tell
the customer that IBM can be depended upon to maintain, expand, and upgrade the
customer's system indefinitely. By contrast, a smaller, unknown, or less well known
computer manufacturer, with no proven track record, or more erratic performance
and growth, has less credibility as a dependable, long-term supplier.

More generally, IBM's stability signals all those qualities that provide customer
satisfaction, for such stability depends on the ability to keep customers satisfied in
the first place. In short, IBM's stability signals dependability in specific and the
ability to satisfy its customers in general. Therefore, as with support and reliability,
stability can reduce perceived purchase risk.

Support, Reliability, and Stability versus Price

The second important observation concerning the purchasing-decision determinancy
rankings are the distinctly low rankings of the price-related variables among the
thirty-three original factors (Table 4.2). More specifically, the majority of factors
ranked in the bottom quintile of all factors (ranks 27 through 33) relate to price or
the cost to the customer: "products are the lowest price," "vendors willingness to
negotiate price," product "offers savings in operator costs," and "vendor offers
large-volume discounts." This is especially striking since it not only means that
support, reliability, and stability are relatively more determinant in purchasing decisions than price, but that price is the least determinant factor relative to the entire
range of factors and thus in an absolute sense. Conversely, support, reliability, and
stability are not simply relatively more determinant than price but are the most
determinant set of factors in the entire range and therefore absolutely most determinant. A similar ranking is found in the clustered data (Table 4.3). "Price flexibility"
is next to last and thus in the bottom quintile of the fourteen macrofactors, while
"absolute price" is in the bottom half of all macroattributes.

These observations provide further empirical support for the theoretical implications of information inadequacy developed in the first part of this thesis. In particular, the data indicate a low customer sensitivity to price in determining purchasing
decisions. The data also indicate a high sensitivity to factors that tend to reduce risk
(customer support and product reliability) and that signal information (firm stability) in determining purchase decisions.

It is logical that the high customer sensitivity to support, reliability, and firm
stability derives from the customer's lack of adequate information to evaluate products on a price/performance basis and consequent desire to reduce purchase risk. In
other words, the data suggest that customers select mainframe computer products
primarily because of perceived augmentation along non–price dimensions and dimensions that specifically signal information and reduce purchase risk. This finding
is significant for market strategy because it points to not only the diminished

importance of price competition and the potential potency of product augmentation but also the importance of behavior that exploits the effects of inadequate information.

Differences in Factors Determining Purchase Decisions
between IBM and Non-IBM Customers

Given the high customer sensitivity to service reliability and firm stability in purchasing decisions and given IBM's market success indicated, in part, by its large market share, it is consistent to conclude that IBM's appeal lies, to a significant degree, in its superior image in the above areas (support, reliability, and stability). Theoretically this position can be tested further by comparing the factor ratings of IBM customers to non-IBM customers. A finding that IBM customers are even more sensitive to these factors than non-IBM customers would provide further support for the idea that IBM's appeal lies in its perceived superiority in these areas.

Figure 4.5 presents determinancy ratings by IBM customers and non-IBM customers for the clustered, macroattributes (the IBM/non-IBM division is not available for the original thirty-three factors). An initial inspection of the data reveals that IBM customers rate "service" and "manufacturers' stability" as more determinant in their decisions than do non-IBM customers. IBM customers further rate "absolute price" and "price flexibility" as less determinant in their purchase decision than non-IBM customers.

The first set of observations as to service and stability, however, are based on ratings differences that are not statistically significant at the 90 percent confidence level (as indicated in the exhibit). The set of observations as to differences in the saliency of price on the other hand are statistically significant. Of the fourteen differences between the IBM and non-IBM ratings, five are statistically significant at the 90 percent confidence level: "absolute price," "price flexibility," "software" (support),[34] "broad [product] line," and "visibility among top management." IBM customers therefore perceive price (absolute price as well as price flexibility) to be less determinant in their purchase decisions than non-IBM customers, while IBM customers are more concerned with software support and a broad product line.

These statistically significant differences are consistent with the idea that IBM's appeal lies in its image in the area of customer support and its orientation toward solving customer problems versus price/performance superiority. The greater importance IBM customers place on product line breadth, in particular, is consistent with the idea that IBM's appeal lies in its image of providing solutions to customer problems insofar as offering a fuller range of products implies a superior ability to find the data processing product that precisely matches the customer's data processing problems.

Information Acquisition and the Dimensions of Product Augmentation

The idea that IBM's augmentation strategy focuses on the enhancement of the "softer," less tangible benefits of support, reliability, and dependability is consistent with the business strategy implications this author developed from the economic framework of information acquisition costs. As Chapter 3 posited, the effectiveness of augmentation is insulating a firm from price competition is related to which

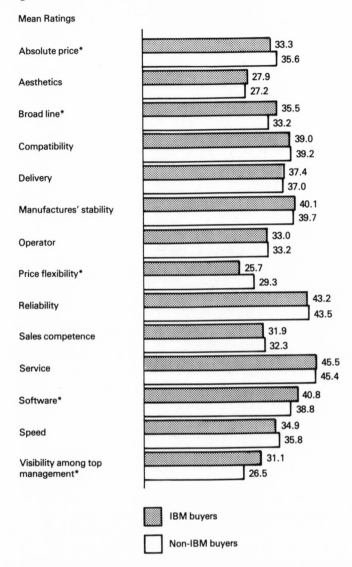

Mean Ratings

	IBM buyers	Non-IBM buyers
Absolute price*	33.3	35.6
Aesthetics	27.9	27.2
Broad line*	35.5	33.2
Compatibility	39.0	39.2
Delivery	37.4	37.0
Manufactures' stability	40.1	39.7
Operator	33.0	33.2
Price flexibility*	25.7	29.3
Reliability	43.2	43.5
Sales competence	31.9	32.3
Service	45.5	45.4
Software*	40.8	38.8
Speed	34.9	35.8
Visibility among top management*	31.1	26.5

▨ IBM buyers

☐ Non-IBM buyers

*Indicates statistically significant difference between the two groups at the 90 percent confidence level.

Figure 4.5. Macroattribute ratings: IBM versus non-IBM customers. *Source:* Moriarty data base.

product attributes are augmented. In particular, product dimensions for which information is harder to obtain, which are also harder to reduce to price/performance measures, are effective areas for augmentation. More specifically, augmentation along experience attributes (for which information can be obtained only after purchases) or credence attributes (for which certain information is not always available even after purchase) is more effective in insulating a firm from price-competitive threats than augmentation along search attributes for which information is available prior to purchase.

Applying these categories to relevant product attributes in the market for mainframe products discussed in this chapter, it is clear that the price-related factors are accurately described as search attributes, while support and reliability seem to better fit the other categories. Indeed, to the extent that support remains unspecified and variable (depending on needs) even after purchase, it may be accurately described as a credence attribute. Since IBM's augmentation strategy focuses on these latter attributes for which information is costly and interproduct comparision difficult, as opposed to price-related variables for which definitive quantifiable information is readily available prior to purchase, IBM's observed behavior is consistent with the theoretical strategy implications drawn from the generic economic framework of information acquisition.

The Role of General Management in Purchasing Decisions

One of the most important observations linking IBM's appeal to the reduction of risk, the signaling of information through reputation, and the strategic use of low buyer sophistication, is the significant difference in the ratings IBM customers and non-IBM customers give to the role of upper-level general management in purchasing decisions. This greater role is confirmed in three different observations. It is first observed in the statistically significant difference in determinancy ratings between IBM and non-IBM customers for the factor termed "visibility among top management" (Figure 4.5). Visibility among top management refers to the degree of recognition the general management of the customer firm has for the mainframe product supplier. This top management recognition is more influential in the purchasing decisions of IBM customers. In particular, the DMUs in IBM customer firms perceive the visibility of IBM among its top managers to be significantly more determinant in their purchasing decision than is the case for supplier visibility among non-IBM customers.

The second place the differentially important role of general management is observed is in the profile data listed in Figure 4.6. Here informal sources of product information are rated for their "importance" in the buying decision. Of all the various sources of informal information about the product to be purchased, IBM buyers rate only one source more important than non-IBM customers: the opinion of top management. This statistically significant difference indicates that IBM buyers place more importance on their top management's opinion of the product than non-IBM buyers in their purchasing decisions.

The differential importance of general management is also observed in profile data on the job levels of those participating in the purchase decision (not exhibited). Employees at various job levels participate in purchasing decisions. The Morierity Data Base reveals that the proportions of these various job levels involved in the buying decision do not vary significantly between IBM customers and non-IBM customers, with one exception: the proportion of "upper management" participating in the decision. In particular, among IBM customers a significantly greater proportion of upper management participates in the purchase decision than is the

Mean of ratings on a scale of 1 to 6

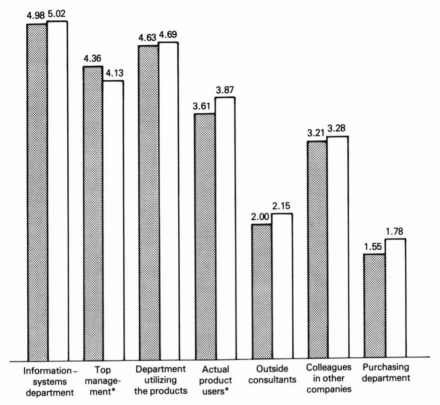

Information– systems department	Top manage- ment*	Department utilizing the products	Actual product users*	Outside consultants	Colleagues in other companies	Purchasing department

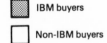

IBM buyers

Non-IBM buyers

*Indicates significant difference

Figure 4.6. Importance of informal information sources: IBM versus non-IBM customers. *Source:* Moriarty Data Base.

case for non-IBM customers. In summary, these three statistical observations reveal that general management's recognition of the supplier, opinions of its products, and participation in the buying process are more significant in the purchasing decisions of IBM customers than non-IBM customers.

The Role of General Management and Its Significance for Reputation
The significantly higher ratings among IBM customers for supplier visibility and the opinion of top management support the idea that an important source of IBM's appeal lies in its reputation. More specifically, the statistical fact that *IBM* DMUs place greater importance on vendor visibility indicates that they perceive IBM to

have a better reputation among their top management. Furthermore, the fact that the DMUs who place greater weight on this top-management opinion are those that select IBM indicates that this top management does, in fact, have higher regard for IBM.

The identification of the importance of reputation in purchasing behavior has significance for explanations of the presence of price premiums or extranormal returns. In particular, because reputation does play a significant role in purchasing behavior in the market for mainframe products in general and even more so for purchasing behavior of IBM customers, the theoretical idea of "returns to reputation" discussed in Chapter 3 is pertinent to interpreting the performance of the market in general and especially pertinent to IBM's performance. More specifically, in light of the significant influence of reputation, any premiums not accounted for by performance differences, or even broader product differences (any "utility" differences), are not necessarily indicative of the unfair rewards of monopoly power but may reflect compensation to the seller for investment in reputation. In this sense, reputation is part of IBM's augmented product and the removal of all product differences would, in theory, not only require the removal of performance differences and "nonhardware service" differences, but also the elimination of differential reputations.

The Role of General Management and Its Significance for Risk Reduction

The greater influence of upper management in buying decisions of DMUs in IBM customer firms, and specifically the first two observations as to the greater importance of vendor visibility and top management opinion, is indicative of the higher importance these DMUs place on pleasing upper management. More precisely, it reflects their greater aversion for purchase risk at the individual level. As Chapter 3 explained, the risk to an individual of making a bad industrial purchasing decision can be much greater than firm-level risk (or *consumer* purchasing risk) because of the relative difference in the possible bad outcomes to the firm versus the individual. The opinion of top management and the image they have of the supplier is important to individual buyers because it reduces the risk to them at a personal level. In other words, the purchase of "IBM" provides a defensible choice for members participating in the purchasing decision making unit, especially against concerns of upper-level management, because of its high visibility, favorable image, and excellent reputation among that management. This interpretation of the data is consistent with the industry folklore among the data processing managers who typically play the most important role in mainframe product purchasing. As the Data Processing Manager and buyer for Conrail, a multimillion dollar IBM account, explained to the author: "No one ever lost their job for buying Big Blue [IBM]."

The Role of General Management and Its Significance
for Exploiting Low Buyer Sophistication

The use of a risk-reduction strategy is a subset of the broader strategic approach that exploits information inadequacy, for it is the information-poor market conditions in this dynamic industry that intensify perceived risk in the first place. Another specific manifestation of this broad strategic approach is the exploitation and manipulation of lack of customer sophistication in making purchasing decisions. The greater role general management plays in purchasing decisions among IBM customers, and specifically the third observation that a significantly greater proportion of general managers participate in the purchasing decision in IBM customers than in non-IBM customers, suggests that IBM's appeal as a purchase choice lies in an advantage along the type of product dimensions that are specifically attractive to general management as opposed to technically expert, purchasing specialists. Since general management has a greater participatory role in the buying decisions of IBM customers, and IBM is the product of choice, it is consistent to conclude that a source of IBM's appeal lies in its relatively greater attractiveness to this less-expert general management group within the customer.

In the terms of market strategy, a source of IBM's marketing effectiveness lies in the targeting of its augmented product to upper-management generalists as opposed to professional purchasing experts. Because such generalist, higher level, decision makers are less capable of, and less comfortable with, selecting products based on technical price/performance evaluations, they are logically more inclined to defer to less sophisticated purchase decision criteria such as the seller's size, stability, and reputation. IBM's targeting of these general managers is thus an attempt to exploit low buyer sophistication.

Furthermore, IBM not only exploits the lack of buyer sophistication that, in part, is caused by the increased role of general management, but may in fact manipulate and, more precisely, exaggerate this lack of buyer sophistication. Such manipulation of buyer sophistication is facilitated through customer management techniques that encourage general management to have an important role in the purchasing decision. The intensification of low buyer sophistication in turn heightens the presence of risk-reducing behavior not only because purchasing is influenced more by upper-level management with less purchasing expertise but also because the lower-level buyers may perceive more personal risk due to greater top-management scrutiny.

This interpretation of the statistical data from the Morierity Data Base is consistent with the position of DeBruicker and Summe based on qualitative data from case study observation. These authors develop a generic hypothesis on market strategy that is said to be supported by observed behavior in IBM. In particular, they conclude that IBM inhibits the evolution of the customer to increasing levels of buyer sophistication through its "account management" marketing technique:

In industries with a brisk rate of new product development, the vendor can slow [customer] evolution by emphasizing account management—strengthening the account management representation and perhaps even adding top management to the team. This strategy is designed to keep purchasing decisions under the regular review of general managers in senior positions, thereby limiting the impact of experienced specialists in important decisions. Its purpose is to sustain the vendor's influence and block competitive inroads into the account. . . . In high tech industries, the fast pace of product and systems development makes account management a feasible strategy, even when customers have achieved some sophistication. . . . IBM used this strategy to great advantage in the computer industry. The company inhibited customers' transitions from inexperienced to experienced buyers by using a multi-level team account management approach. Including high-level managers . . . in the decision making unit, ensured the continued involvement of inexperienced generalists in the decision making process. These moves enabled IBM to retain most of its customer base over the long term.[35]

What the authors of the quotation do not point out, however, is that the reduction of sophistication in buying decisions is not only facilitated by encouraging of the active, ongoing participation of general management in the buying decision, but it is also facilitated by encouraging the general management to establish preset purchasing guidelines. Such guidelines free general management from the need for active participation yet at the same time impose general management preferences and purchasing criteria on the decisions of purchasing specialists. The consequent institutionalized bias for the vendor that appeals to general management—IBM—can be so strong as to completely displace a competitive buying process. In such cases, as one expert witness testifying against IBM explains, "the customer was being directed by a corporate data processing policy to buy from IBM only."[36] Rigid corporate directives established by general management can be viewed as, in a sense, the ultimate desired conclusion of a market strategy designed to keep purchasing decisions with general managers in senior positions. Moreover, it is not difficult to justify the need for such rigid purchasing guidelines because "large, multidivisional companies often sought to have each of their individual data processing centers standardize on systems from a single vendor to facilitate easy communication of data and application programs between several systems and to avoid the redundant training of personnel."[37]

Note carefully that the concept of exploiting low buyer sophistication as used in this study does not mean betraying a buyer's trust by not providing a customer with what he expects as a result of the information signaled by reputation. Rather it refers to the use of low buyer sophistication as a vehicle for product differentiation that insulates the supplier from competitive threats. In fact, the ability of IBM to sustain its reputation indicates that such customer trust is not betrayed. Indeed the maintenance of reputation in light of low buyer sophistication points to the applicability of one of the theoretical implications of the asymmetric information framework discussed in Chapter 3. In particular, the presence of higher-than-normal returns, after all product utility differences are removed, are not necessarily the result of the unfair, excess extraction of customer revenue stemming from monopoly power and, in fact, may reflect the opportunity cost to IBM of forgoing a more profitable short-term strategy of "exploiting buyer ignorance" (betraying buyers trust) in favor of a long-term strategy of maintaining superior product benefits.

Two Sides of IBM's Product Augmentation Strategy: Product Bundling and Customer Account Management

Juxtaposing the different logic behind two marketing techniques used by IBM discussed in this chapter, account management and product bundling, shows the applicability of the theoretical idea of two sides to each strategy type developed in Chapter 3. In particular, one theoretical implication of the notion of the function of markets as a matching process was the identification and clarification of two sides to product augmentation, to product segmentation, and to price segmentation. More specifically, it was theoretically illustrated how each of these behaviors could entail either altering the perceived product to fit customer wants or altering customer wants to fit the perceived product.

This distinction is applicable to actual product augmenting behavior discussed in this chapter. Product bundling allows IBM to attain a differential product advantage by actually altering the nature of the product in that it bundles together hardware, software, and unspecified support. Account management on the other hand assumes a product difference (unspecified support and, more generally, IBM's risk-reducing image) and manipulates the customer's desire for that product. Specifically, as discussed, account management exaggerates low buyer sophistication and thus the desire for a "safe bet" purchase choice. Hence, in the first case, IBM alters perceived products to better fit customer wants and, in the second case, IBM alters customer wants to better fit the perceived product. In both cases the price sensitivity-reducing effect of product differentiation is achieved.

Risk Reduction: Other Evidence

Customer Risk Ratings
Other customer profile information from the Morierity Data Base, directly addressing the customer's perception of purchase risk, reveals important differences between IBM and non-IBM customers. Although the precise magnitude of the actual importance ratings was not available from the data base in all cases, the fact of occurence of significant differences was in all cases available. Specific factors that IBM customers rated significantly more important in purchase decisions in the profile data areas of "perceived risk" and "decision conflict" are listed below:

- "General risk level to the buyer's firm"
- "General risk level to the buyer personally"
- "Risk of inadequate reliability to the buyer's firm"
- "Risk of inadequate reliability to the buyer personally"
- "Conflict in purchase decision"
- "Need to feel confident about products purchased"

IBM buyers therefore perceived the general level of risk, both to the firm and to themselves, as more important in their purchase decision than non-IBM buyers. The specific risk of inadequate reliability was also seen to be more important among IBM customers both at the firm and individual levels. IBM buyers also felt their decision entailed more conflict and, relatedly, placed more importance on feelings of confidence about the product purchased. The greater perceived risk among IBM buyers, both general and specific reliability risk, as well as the greater perceived decision conflict and need for decision confidence, directly support the idea that a source of IBM's market effectiveness lies in its ability to reduce the customer's perception of purchase risk.

Customer Size
Descriptive customer profile data (as opposed to profile data summarizing customer perceptions) and specifically data on customer size, buying decision size, and purchase type lends further support to the important strategic role of customer risk

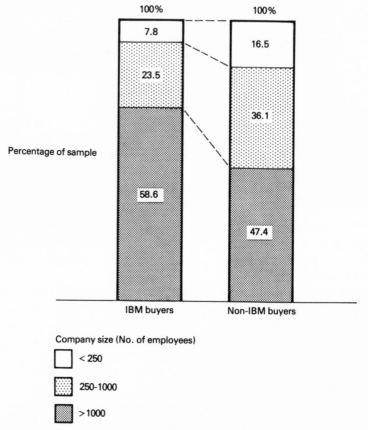

Figure 4.7. Customer size: IBM versus non-IBM buyers. *Source:* Moriarty data base.

reduction in IBM's market effectiveness. Data on customer size seems consistent with the notion of IBM's low-risk appeal. The data indicate company size is positively correlated with the decision to purchase IBM. As Figure 4.7 shows, while small firms favor non-IBM products 2 to 1 (dark grey), the dominant choice of large firms is IBM (light grey).

The preference for IBM among large firms may, in part, by accounted for by the different degrees of discrepancy between firm risk and personal risk at large versus small firms. In small firms (under 250 employees) the actual individual purchaser may also be among the general management. Firm risk and individual risk are likely to be balanced if not one in the same, for career success is more closely aligned with firm success. In particular, the low-risk appeal of IBM will be balanced against the efficiency need to keep purchase costs down. In large firms, specialized computer purchasing departments, not part of general management, are likely to identify less with firm goals. Purchase decisions will thus be more weighted toward consider-ations of individual-level risk. Recalling the discussion of risk in Chapter 3, to the extent that individual risk is likely to be more extreme, large customers may be more inclined to buy products known for being safe career bets rather than balancing career safety with the firm goal of efficiency and therefore a low purchase price. Hence both the generic idea as to the difference between firm risk and individual risk, as well as the specific notion as to the role of risk reduction in IBM's strategy, seem consistent with the pattern observed in the customer size data indicating large customers tend to select the low-risk vendor, IBM, while small customers prefer the more price-competitive IBM rivals.

Purchase Size

Data on the average size of the buying decision lends further support to the notion that low-risk appeal plays a significant role in IBM's market effectiveness. Figure 4.8 shows that the typical purchase among IBM customers involves fewer units but a large financial outlay. More specifically, the average IBM customer purchase is less than one quarter of the number of products purchased in an average non-IBM customer purchase. On the other hand, the typical IBM customer purchase *costs* well over triple that of the typical non-IBM customer purchase. This means that purchases by IBM customers tend to involve a higher level of financial commitment both on a per product basis and with respect to the total units purchased.

Since, as Fisher explains, purchase risk in the mainframe industry (as in other industries) is related to the "financial risk in terms of opportunity foregone"[38] and

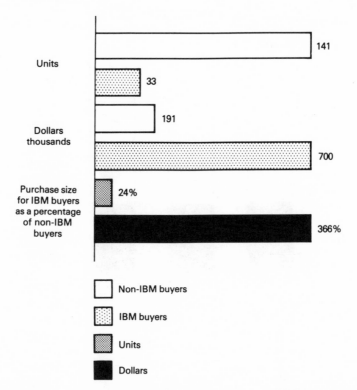

Figure 4.8. Average size of buying decision: IBM versus non-IBM customers. *Source:* Moriarty data base.

since, statistically, IBM customer purchases involve a greater commitment of finan-cial resources, IBM customers will tend to perceive higher levels of purchase risk. The data is thus consistent with the idea of IBM as the low-risk choice in that the customers who perceive higher risk as a result of having to make more costly purchases, also are those who tend to select IBM.

Purchase Type

Descriptive profile data on the type of purchase offers further support for the idea of IBM as the low-risk vendor. The data base classifies purchases into three types: "pilot" purchases for trial runs prior to major purchases; "implementation" pur-chases for new customer requirements due to, for instance, growth or new applica-tions; and "replacement" purchases to sustain existing system capacities. While there is no significant difference in these last two types of purchase between IBM and non-IBM customers, pilot purchases do differ significantly. In particular, as Figure 4.9 shows, of the 12 percent of purchases across all customers that involved pilot purchases, IBM customers used pilots in only about half the proportion to their total purchases than non-IBM customers did (8.5 versus 14.5 percent). The greater use of pilots by non-IBM customers, and the lesser use of pilots by IBM customers, is consistent with the idea of IBM as a low-risk choice.

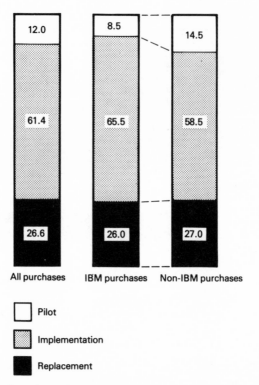

Figure 4.9. Type of purchase: IBM customers versus non-IBM customers percent of pur-chases. *Source:* Moriarty data base.

A pilot program is a means for the buyer to test the adequacy and effectiveness of a product before a major purchase is approved. Once products are judged adequate or effective, purchasers are freed to focus on other criteria, such as volume discounts, in making the major purchase. In short, pilots reduce the risk of inadequacy *prior* to the major purchase. The greater use of pilots prior to major purchases by non-IBM customers is therefore consistent with their greater sensitivity to price factors and lower sensitivity to risk considerations during major purchases. The lesser use of pilots among IBM customers, and thus the fewer occasions on which risk is reduced prior to purchase, is consistent with the higher levels of risk perceived by IBM customers. The pilot data is consistent with the idea of IBM as the low-risk vendor because customers with lower pilot usage, and hence greater purchase risk, are the same customers that select IBM. In summary, the statistical facts describing customer size, purchase size, purchase type, and specifically that IBM customers tend to be larger in size, make more costly purchases, and use less pilots is consistent with notion that risk reduction plays an important strategic role in creating and sustaining IBM's advantageous market posture.

CHAPTER SUMMARY

This chapter illustrated how the context of mainframe product market behavior intensifies the danger and diminishes the effectiveness of price competitive behavior, particularly to IBM, while at the same time provides conditions appropriate for the use of risk-reducing, product-dimensional behavior. The idea that such risk-reducing behavior was in fact present and did serve as an important source of IBM's market effectiveness received empirical support through a variety of observations. Empirical economic research, customer survey data, IBM's own internal product assessments, as well as testimony by IBM executives and industry observations by others all suggested that, even in the heyday of IBM's technological innovativeness, its advantage could not be explained by superior price/performance and pointed toward IBM's image of offering more in terms of "nonhardware services" as a source of its market effectiveness.

This idea was supported and extended with detailed statistical data on purchasing behavior. The data indicated that customers in general purchase mainframe products because of risk-reducing product attributes (customer support, product reliability, firm stability). Conversely, price-related factors were seen to be among the least significant in determining purchasing behavior.

The data also had significance for understanding specifically IBM's behavior. It established the important empirical finding that those who select IBM are also those whose purchasing situation (greater general management influence and participation), purchasing behavior (larger buys among larger customers entailing fewer pilots), purchasing preferences (greater importance of customer support, lesser importance of price), and actual ratings of risk indicate a greater need for reduced purchase risk. The fact that these customers with higher perceived purchase risk are also those customers who select IBM indicates that product-dimensional, risk-reducing behavior is an important source of customer effectiveness for IBM.

Through the course of this analysis, theoretical economic ideas (asymmetric information, information acquisition, and returns to reputation), as well as central ideas of this study were shown to be applicable to, and consistent with, the emerging interpretation of IBM's behavior. Furthermore, this interpretation based on statistical compilations of customer responses was also supported by actual observations as to the special features of IBM's market behavior. In particular, the marketing techniques of bundling and account management as practiced by IBM, as well as the extensive use of non–sales field personnel, were seen to serve as tools of product dimensional strategy that exploit the related effects of information inadequacy and that specifically exploit low buyer sophistication, high receptivity to signals, and most importantly, the customer's perception of purchase risk.

NOTES

1. Levitt, *The Marketing Mode*, p. 3.

2. S. J. Ippolito, "Measure for Counter-Measure," *Datamation* (February, 1979).

3. Other reasons for avoiding price-cutting activity by the dominant firm obviously include potential charges of "predatory pricing" by antitrust authorities as well as the less obvious cost structure differences between IBM and much smaller firms. Technology-driven direct cost reductions can provide greater strategic advantage to smaller firms whose total costs contain a higher portion of direct costs.

4. Geruson, "Elements of Strategy in IBM's Price and Product Behaviour" (Oxford University M.Phil. thesis, 1982), pp. 59–100.

5. W. H. Davidson, *"The Amazing Race: Winning the Technorivalry With Japan"* (New York, John Wiley and Sons, 1984), p. 3.

6. R. Sobel, *IBM vs. Japan: The Struggle for the Future* (New York, Stein and Day, 1986), p. 243.

7. R. T. DeLamarter, *Big Blue: IBM's Use and Abuse of Power* (New York, Dodd, Mead, 1986), p. 67.

8. *United States v. IBM*, Civil Action No. 69, Civ. 200 (D.N.E.), United States District Court for the Southern District of New York, filed 17 January, 1969, dismissed 8 January 1981. (PX means Plaintiff Exhibit; DX means Defense Exhibit; Ex. means joint exhibit; Tr. refers to the transcript page of sworn witness testimony in the trial record.) Withington of A. D. Little, Tr. 58533, 58463, 58797–98, 58830, 58423, 56566; Beard of RCA, Tr. 8493–96; McCollister of RCA, Tr. 9269–73; Bloch of Honeywell, Tr. 7599–7601, 7596–97; Hangen of NCR 6350–52; McDonald of Sperry-Univac, Tr. 2883–86, 4150–56, 4178–79, 4217–19; Norris of CDC, Tr. 5653–54; Lacey of CDC, Tr. 6570–73; Palevsky of SDS 3149–50, 3165, 3176; and Wright of Xerox, Tr. 12182–84.

9. S. Rosen, "Hedonic Prices and Implicit Markets: Product Differentiation in Pure Competition," *Journal of Political Economy*, 82, No. 1 (January–February, 1974), p. 34.

10. B. T. Ratchford, G. T. Ford, "A Study of Prices and Market Shares in the Computer Mainframe Industry," *Journal of Business*, 49 (1976), pp. 194–218.

11. Ibid., p. 218.

12. Ibid., p. 217.

13. Ibid., pp. 217–218.

14. G. W. Brock, "A Study of Prices and Market Shares in the Computer Mainframe Industry: Comment," *Journal of Business*, 49 (1979). Brock pointed out and, in fact, analyzed the difficulty of evaluating the relative prices of computers in 1975 before any of the above studies were undertaken. See Brock, *The U.S. Computer Industry* (Cambridge, MA, Ballinger, 1975), pp. 44–49 and 37–42.

15. B. T. Ratchford, G. T. Ford, "A Study of Prices and Market Shares in the Computer Mainframe Industry: A Reply," *Journal of Business* (1979), pp. 125–134.

16. R. Michaels, "Hedonic Prices and the Structure of the Digital Computer Industry," *The Journal of Industry Economics*, 27, No. 3 (1979), p. 273.

17. Levitt, *Marketing for Business Growth*, p. 10.

18. T. Levitt, *The Marketing Imagination,* p. 74.

19. T. Levitt, Marketing and the Corporate Purpose in *Changing Marketing Strategies in a New Economy,* edited by J. Backman, J. Czepiel (Indianapolis, Bobbs-Merrill, 1977), p. 13.

20. *U.S. v. IBM,* R. Block, Tr. 7753–54; see also McCollister Tr. 11041–43; Norris Tr. 6058; Welke Tr. 17380–81, 17343–46; and PX 1096, p. 1.

21. Defense Exhibit 13849, p. 27.

22. T. Levitt, *Marketing for Business Growth,* p. 10.

23. *U.S. v. IBM,* Welke Tr. 19225–26; 19228.

24. *U.S. v. IBM,* PX 1256; DX 14504.

25. Brock, *The U.S. Computer Industry,* p. 48.

26. F. S. DeBruicker, G. L. Summe, "Make Sure Your Customers Keep Coming Back," *Harvard Business Review,* 85, No. 1 (January–February, 1985), p. 93.

27. Note carefully that this generic job position is widely held throughout IBM although the name it takes has varied with time and place. At different times and in different IBM divisions, this group of field personnel have been variously called "Customer Engineers" or "CEs," "Customer Service Representatives" or "CSRs," "Field Service Representatives" or "FSRs," and "Systems Engineers" or "SEs."

28. *U.S. v. IBM,* DX 52, p. 1.

29. *U.S. v. IBM,* DX 1525.

30. *U.S. v. IBM,* DX 14295, p. 44.

31. *U.S. v. IBM,* DX 13824, p. 2.

32. DeLamarter, *Big Blue,* p. 81.

33. All the *statistical* data used in this section are taken from what is hereafter termed the "Morierity Data Base." See Morierity, *Industrial Buying Behavior.*

34. The macroattribute "software" represents software support. See Figure 4.4 for the three original factors from which it is derived.

35. DeBruicker, Summe, "Make Sure Your Customers Keep Coming Back," p. 96.

36. *U.S. v. IBM,* PX 4060.

37. *U.S. v. IBM,* PX 2508A.

38. F. M. Fisher, J. W. McKie, R. D. Manke, *IBM and the U.S. vs. Data Processing Industry: An Economic History* (New York, Praeger, 1983), p. 19.

5

Price Segmentation by IBM in the Market for Mainframe Computer Products

IBM . . . is dependent on . . . price control. IBM controls (price) by various means: timing of new technological insertation; functional pricing, . . . refusal to market surplus used equipment; refusal to discount for age or for quantity; strategic location of function in boxes; solution selling rather than hardware selling. . . . The key underpinnings to our control of price are . . . not well understood by IBM management. Our price control has been sufficiently absolute to render unnecessary direct management involvement in the means.

Internal confidential memo from IBM
financial executive Hillary Faw to
then chief marketing executive
(soon-to-be IBM CEO) Frank Carey[1]

This chapter presents detailed evidence for both an emergent and intended strategy of segmented pricing: the exploitation of customer heterogeneity as to price sensitivity. As explained in the Introduction, "emergent strategy" is revealed by discerning patterns in observed firm behavior. For purposes of strategy research, this is generally the only methodologically practicable definition of strategy, and it is this method that provides the strongest evidence for segmented pricing. Throughout the presentation, this evidence is supplemented with quotations from sworn court testimony by IBM executives and confidential IBM documents publicly disclosed as a result of various private and public law suits brought against IBM and especially the most recent suit brought by the U.S. Justice Department. This additional data provides an unusual opportunity for observing intended strategy.

From the outset it must be understood that the objective of this chapter is not to pass judgment on the legality of IBM's behavior; the objective, rather, is to provide evidence for price segmentation in the economic sense. And, in fact, contrary to what might be implied by the failure of Justice Department lawyers to establish that IBM violated discriminatory-pricing laws in the long and costly court battle that was eventually dismissed as "without merit," careful review of IBM's behavior reveals a distinct pattern of price segmentation.

Apart from the different requirements between legal evidence and evidence for business strategy research, perhaps one reason for the failure of the Justice Department to prove its case lies in the exceptionally detailed level of observation required before the pattern emerges. More specifically, this author has found that observation of product and price behavior at the ordinary levels of market analysis—the product line, the product series, or the product model—often reveal no particular pattern of price segmentation. In some cases it is only when IBM's pricing behavior is reviewed at the submodel level, that the presence of price segmentation is clear.

The data used in this chapter are historical data in the sense that the data span a range of product generations in IBM's history and that the data do not extend beyond the mid-1970s (because, unfortunately, court records do not disclose needed "IBM-confidential" data later than that time). The historical nature of the data, however, should not be misinterpreted as indicating this chapter is a business case history; it is not. The data, although historical in nature, are *not* used to show the chronological evolution of IBM's behavior; rather, the data serve as evidence for price segmentation. The order of the presentation of the data corresponds to the order of the analyses, not when the observed behavior occurred in time. For the reader who wishes to clarify the dates of the behavior discussed, a reference is provided in Table 5.1. Note also that performance data on IBM and its major competitors for the later part of this period are provided in Appendix C.

Beyond supporting the general point that price segmentation can be an important strategic tool, the activities reviewed here support other subpoints made in the price segmentation section of Chapter 2. As discussed in that chapter, the practical application and thus strategic impact of segmented pricing has been viewed as limited because of legal proscriptions, methodological problems, and operational

Table 5.1 Dates of Important Legal Developments and of Product Data Used

1936	First Consent Decree handed down
1947	Investigation begins leading to Second Consent Decree
1952	Charges filed leading to Second Consent Decree
1956	Second Consent Decree
1959	IBM 709 Series
1960	IBM 7070 Series
1961	IBM 1620 Series
1962	IBM 1410 Series
1963	IBM 1401 Series (functional pricing data)
1964	Preliminary inquiries made in *U.S. v. IBM* case
1964	IBM 7010 Series
1965	IBM 2400 Series
1965	IBM 1401 Series (metering data)
1966	IBM 1130 Series
1966	IBM 360 Series (sub-model data)
1967	IBM 360 Series (customer support data)
1969	Charges filed in *U.S. v. IBM* case
1969	IBM 2300 Series
1972	IBM 370 Series
1973	Out of court settlement in *Telex v. IBM* case
1975	Trial begins in *U.S. v. IBM* case
1978	Directed verdict for IBM in *Memorex v. IBM* case
1982	*U.S. v. IBM* case withdrawn because it is "without merit"

difficulties in identifying and measuring the price elasticities assumed to be required for implementing price segmentation. Moreover, to the extent that an important role for price-segmenting behavior was acknowledged, it applied to consumer markets.

Contrary to this perspective, this chapter indicates that price segmentation plays an important strategic role in an industrial market—despite legal prohibitions. Furthermore, this price-segmenting activity is shown to exist in the absence of any attempt by IBM to identify customer groups by measured elasticity difference as commonly done in consumer markets. This chapter further points to the practical application and strategic importance of the more recent theoretical contributions to price-segmentation logic in the areas of tied sales, metering, and product bundling. It also reveals other forms of behavior that extract consumer surplus while avoiding explicitly discriminatory pricing (as do tied sales, metering, and bundling) but in ways not considered in such generic formal analyses.

FUNCTIONAL PRICING IN THE MAINFRAME MARKET

This major section presents evidence for a particular form of segmented pricing. What IBM labels "functional pricing" is here shown to be segmented pricing in the less strict definition of the pricing of similar products such that price differences are not proportional to differences in cost. Note that although this definition is not the definition of pure price discrimination in the sense of price differences for identical products, it is consistent with contemporary economic theory, legislation, and most importantly the business strategy idea of exploiting customer heterogeneity with respect to reservation prices.

Demonstrating price-segmenting behavior among identical products is straightforward; only pricing data is required. Demonstrating the presence of price segmentation in the less strict sense is more complex because it also requires cost data and thus profit data by product. It is also more difficult because profit data are harder to obtain, especially on a product-by-product basis. In this case it would initially seem extraordinarily difficult to show because, as stated, the pattern sometimes exists only at the submodel level.

Publicly disclosed court documents, however, reveal profit data at these very detailed levels sufficient to establish patterns of price segmentation. The following subsections of this major chapter section present data for five IBM product series: the 1401 CPU series, the commercial 360 CPU series, the noncommercial 360 CPU series, the early 2400 tape drive series, and the later 2400 tape drive series. This data show financial performance, not only simply for the series or even the models within the series, but for each submodel within each model. It is at this level—differences in the pricing of submodels—that price segmentation is revealed. In each of the five cases, IBM used product differences at the submodel level as a vehicle for charging different prices. But profit (and cost) data reveal that these price differentials were not related to cost differentials, and patterns in the data in fact suggest a systematic program of price segmentation.

Memory-Based Pricing of the IBM 1401 Series CPUs

The IBM 1401 mainframe came in different models, each with a different capacity to use a selection of special, separately priced mainframe peripheral products, as well as several "optional features." Optional features included such items as special circuits to speed multiplication and facilities for connecting advanced peripheral devices. The 1401 model A, the simplest CPU, for instance, was able to use only four of ten special features; the 1401 model C, on the other hand, included five of these features in the purchase price and was capable of using all ten if the user was willing to pay the additional charge. Each model came in three submodels distinguished by the amount of memory included in the purchase price. The memory-capacity sizes offered were 1400, 2000, or 4000 cells of magnetic core memory respectively, labeled size 1, size 2, and size 3 by IBM. It was these submodel differences that served as the basis for segmented pricing.

Charging different prices for machines distinguished by memory capacity is not price segmentation per se. But IBM charged prices for each memory size that were disproportionate to cost differences. This is clearly seen in Figure 5.1, which shows that for each model IBM received three different levels of profit margin. The relationship of price to cost is much different for type-3 processors than type-1 processors. In particular, on type-3 processors, with margins close to 40 percent, IBM earned profits at a rate one-third higher than on type-1 processors, which had margins of about 30 percent.

One can argue that IBM was merely charging on the basis of performance rather than cost (although this does not diminish the case for price segmentation). In fact, however, in this instance, price differentials are not only not related to costs, but are not related to performance differences either.

Table 5.2A gives the prices of the 1401 series by model and submodel obtained directly from IBM price lists. As calculated in B of the same exhibit, the price differential between submodel 1 and submodel 2 is the same for all models ($800). The price differential between submodel 2 and submodel 3 is also the same for all models ($1000), and therefore the price differential between submodel 1 and submodel 3 is the same for all models ($1900).

By relating these three price differentials to the memory size differences associated with each, we can test if IBM is basing price on memory performance. In fact, as Figure 5.2 reveals, the ratio of incremental price to incremental memory capacity (represented by the slope of the lines) is different for all three possible price differentials. For instance, a customer moving form size 1 to size 2 is paying $1330 per 1000 memory cells ($800/600 cells), a customer moving from size 2 to size 3 is paying $550 per 1000 cells ($1100/2000 cells), and a customer upgrading directly from size 1 to size 3 is paying $731 per 1000 cells ($1900/2600 cells).

If per unit cost data ($212 per 1000 cells) are superimposed on the incremental price/performance diagram, it becomes clear that IBM's incremental pricing is also not related to cost. As Figure 5.2 indicates, the profit margins on upgrades varied from 61 cents on the dollar to 84 cents on the dollar, depending on the type of upgrade.

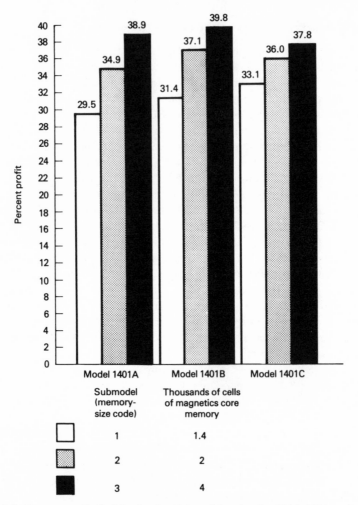

Figure 5.1. Percentage Profit Margins by Submodel series for IBM 1401 series CPUs. *Source: U.S. v. IBM,* Plaintiff Exhibit 1001A, p. 14.

The differentially higher profit rates associated with machines with greater memory were amplified by the tendency for large memory customers (model C customers for instance) to use more add-on features. These features were "functionally priced" to earn as much as 86 cents per dollar of revenue and, on average, across the 1401 series systems, to yield 58 percent margins. The rationale for this feature pricing is made quite explicit by Faw, the former head of IBM's pricing methods area, who wrote that "optional feature prices were established so as to yield substantially higher profit margins than the basic machines."[2] In particular, "features were priced to yield an average of 56 percent profit as compared with an average of 36.8 percent for the basic machines."

Table 5.2 Pricing of the IBM 1401 Series CPUs

	A		
	Submodel 1	Submodel 2	Submodel 3
Model A	$70,500	$71,300	$72,400
Model B	$78,000	$78,000	$79,900
Model C	$105,400	$106,200	$107,300
Memory cells	1400 cells	2000 cells	4000 cells

	B		
	Delta (2-1)	Delta (3-2)	Delta (3-1)
Model A	$800	$1100	$1900
Model B	$800	$1100	$1900
Model C	$800	$1100	$1900
Memory cells	600 cells	2000 cells	2600 cells

Sources: IBM price lists (publicly available). The data in Table B are derived from Table A; the data from Table A are also found in the larger set of data in De Lamarter, *Big Blue*, p. 50.

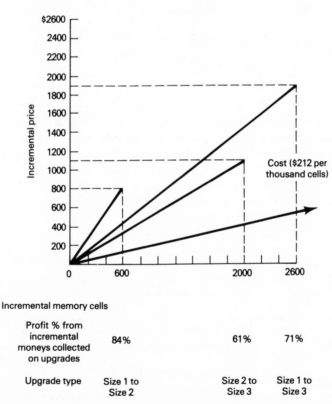

Incremental memory cells

Profit % from incremental moneys collected on upgrades	84%		61%	71%
Upgrade type	Size 1 to Size 2		Size 2 to Size 3	Size 1 to Size 3

Figure 5.2. Profitability of submodel upgrades for IBM 1401 series CPUs. *Source:* Calculated from *U.S. v. IBM,* Plaintiff Exhibit 1001A, p. 14.

Memory-Based Pricing of the IBM 360 Series CPUs

An even more strikingly discriminatory pricing pattern can be seen at the submodel level of the IBM 360 CPU series. As with the 1401 mainframe series, the 360 series came in a variety of models distinguished by performance and functionality. The five main commercial 360 processors were the models 20, 30, 40, 50, and 65. The larger the model number, the larger the number of peripheral devices (disk drives, tape drives, printers, and remote communications equipment) with which each could be equipped. Like the 1401 processors, the 360 processors were available in several submodels according to memory size. But unlike the 1401, each 360 model was limited to a range of memory sizes based on its speed. The memory sizes available for each 360 model are listed in Table 5.3, which gives sizes in bytes (the standard measure of memory capacity for 360-generation technology) and not the number of memory cells (as in the case of the 1401).

The right-most column of Table 5.4 reveals that, for each model, profit margins varied dramatically. The extreme range of margin variability is expressed diagrammatically in Figure 5.3. What immediately becomes clear from the diagram is that for all models, with the exception of the 360 model 50, IBM priced its low-end submodels below costs, while the top submodel of each model was typically priced at the 30 to 40 percent margin level.

By accumulating the total revenue and the total profits across all models, an aggregate picture of the profit implications of this pricing program emerges. Figure 5.4, based on total commercial 360 revenues and profits, shows the accumulated margins for models 20 through 65 by submodel type as determined by relative memory capacity within each model category. For all submodels possessing the largest memory size available for that model, margins exceeded 37 percent; while for all those submodels with the smallest available memory configuration, IBM lost over 17 cents on the dollar. IBM received successively greater return on sales the closer the submodel was to the maximum memory configuration limit for that model.

The noncommercial market also shows the same pattern of intra-submodel price

Table 5.3 Submodel Memory Sizes for IBM 360 Series Commercial CPUs

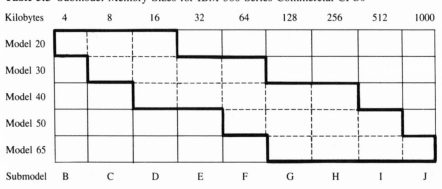

Kilobytes	4	8	16	32	64	128	256	512	1000
Model 20									
Model 30									
Model 40									
Model 50									
Model 65									
Submodel	B	C	D	E	F	G	H	I	J

Source: U.S. v. IBM, Plaintiff Exhibit 1962 A; data also available in De Lamarter, *Big Blue,* p. 76.

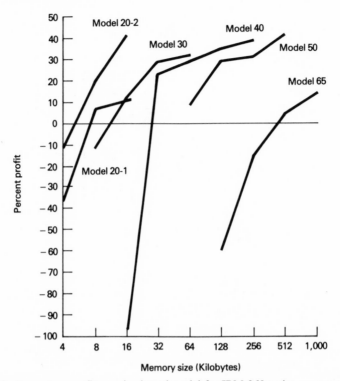

Figure 5.3. Percentage profit margins by submodel for IBM 360 series commercial CPUs. *Source:* IBM Blue Letters and *U.S. v. IBM*, Plaintiff Exhibit 1962A.

segmentation, with the highest profits being extracted from the high-end submodels and with losses being sacrificed in the low-end submodel market (Table 5.5). Large overall losses on the noncommercial systems reflect further price-segmenting behavior at a more macro level and, in particular, across the commercial and noncommercial markets. As the bottom lines of Table 5.6 show, the actual margins for the commercial market were more than double the margins in the noncommercial market. The lower margins were accepted and the large losses on the low-end noncommercial market mainframes were tolerated because, as will be discussed in Chapter 6, the purchasing example set by scientific customers in the noncommercial market had a strong influence on purchasing decisions in the commercial market.[3] Given the relatively small size of the noncommercial market and therefore the small effect its profitability has an overall profitability (as Table 5.6 shows), the losses and lower margins in this market are a small price to pay for sustaining IBM's reputation as an industry leader and the "choice of experts"—a reputation important to sales in the much larger commercial market.

Transfer Rate-Based Pricing of the IBM 2401 Series Disk Drive

A similar pattern is apparent in IBM's pricing of the IBM 2400 series tape drive, a mainframe peripheral product that attached to the 360 series CPU. Like IBM pro-

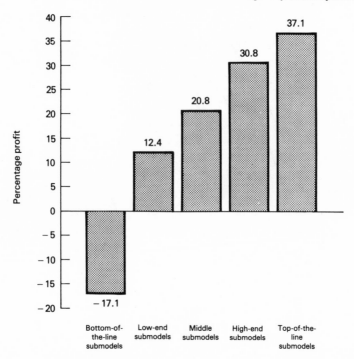

Figure 5.4. Profitability by submodel type for IBM 360 series commercial CPUs. (See Appendix D for submodels included in each category.) *Source:* Calculated from IBM Blue Letters and *U.S. v. IBM*, Plaintiff Exhibit 1962A.

cessors, the 2400 series tape drive came in several models and submodels. Submodels were distinguished on the basis of varying data transfer rates. Transfer rates measure how many 8-bit characters move to and from the CPU in a second and depend on the density at which data are stored on the tape as well as the rapidity with which the drive moves the data past the read/write head.

As the two left-most columns of Table 5.7 show, there were three basic models, all of which were available in a total of six submodel transfer rate speeds. Initially each model came in only three submodels: 1, 2, and 3 representing 30, 60, and 90 thousand 8-bit characters per second transfer rates respectively. Later IBM introduced new generation technology submodels 4, 5, and 6 representing 60, 120, and 180 thousand 8-bit characters per second. Within both the early and later generations segmented pricing is clearly discernible. Figure 5.5 depicts the submodel profit percentages associated with each generation. In the early generation the bottom-of-the-line submodel within each model (submodel number one) was priced to just break even, while the top submodel for each model (submodel number three) earned IBM from 20 cents to 40 cents on the dollar.

The same general pattern is seen in the later generation; the range of margin variability is similarly wide. Later generation model 2402 customers using the top-of-the-line submodel (2402-6) pay IBM over 40 cents per dollar in profit while 2402 customers using the lower end submodel (2402-4) are covering IBM's costs by a margin less than a third of this size. In both the early and later generations, price differences between submodels are substantially disproportional to differences in cost.

The Strategic Logic and Implications of IBM's Functional Pricing

The data analysed in this section strongly suggest the presence of price segmentation. The differences in products at the submodel level and, in particular, differences in memory or transfer rates served, in a sense, as an excuse to charge different prices that were disproportionate to cost differences. Such a pricing program has no intended relation to costs; rather it is based on a strategy of expanding the customer base (increasing the width of the cake) by attracting highly price-sensitive custom-

Table 5.4 Revenue and Profitability by Submodel for IBM 360 Series Commercial CPUs* ($ millions)

Model and Submodel	Total Revenues	Total Profits	Margin (%)
360/20-1		$-5.8	-37.2
B	$15.6	9.0	6.8
C	133.2	1.0	11.9
D	8.1		
360/20-2			
B	6.4	-.8	-12.3
C	74.5	15.2	20.4
D	230.8	94.6	41.1
360/30			
C	44.9	-5.2	-11.6
D	279.0	33.6	12.0
E	555.5	160.1	28.8
F	417.7	134.7	32.2
360/40			
D	.6	-.6	-96.8
E	30.8	7.0	22.7
F	167.3	53.9	28.8
G	441.0	153.2	34.7
H	194.6	75.9	39.0
360/50			
F	4.9	.4	8.4
G	98.5	28.9	28.9
H	288.0	88.9	30.9
I	68.0	28.2	41.5
360/65			
G	2.9	-1.7	-59.3
H	79.8	-12.0	-15.1
I	94.1	4.3	4.6
J	8.4	1.2	13.9

Source: IBM Blue Letters and *U.S. v. IBM,* Plaintiff Exhibit 1962A; these data are also available from the larger set of data in De Lamarter, *Big Blue,* p.77.

*Note: Not all submodels included.

Table 5.5 Profitability by Submodel for IBM 360 Series Noncommercial CPUs* ($ millions)

Model and Submodel	Total Revenues	Total Profits	Margin (%)
360/44			
E	$5.63	$-1.53	$-27.1
F	19.63	.58	2.3
G	28.51	.67	2.3
H	25.89	5.39	20.8
360/67			
1	29.56	-46.71	-158.0
2	75.96	-35.76	47.1
360/75			
H	6.28	-6.61	-105.2
I	32.35	-20.44	-63.2
J	14.22	-8.79	-61.8

Source: U.S. v. IBM, Plaintiff Exhibit 1962A; data also available in De Lamarter, *Big Blue,* p. 91.

*Note: Not all submodels included.

ers with extremely low prices and extracting the greatest revenue possible from existing less price-sensitive customers (increasing the height of the cake by adding higher and higher multiple-tiered layers).

This conclusion, based on observed pricing behavior, is reinforced by the sworn court testimony of former IBM pricing executive Northrop, who states that IBM prices its mainframes "according to the value of the product to the user and not the costs to IBM of producing the product."[4] Faw discloses that IBM's "boxes were priced and marketed on the concept of displaceable cost: discrete measurable value [to customers, the] determinable cost of customers' current administrative methods . . . displaced rather than cost of machine-service package being offered."[5]

The key to understanding how the differential pricing of submodels allowed IBM

Table 5.6 Total Revenue and Profitability for IBM 360 Series CPUs ($ millions)

Commercial Models:	Total Revenues	Total Profits	Margin (%)
20	$ 1,644	$	29.9
30	3,914	1,241	31.7
40	2,027	709	35.0
50	926	307	33.2
65	776	185	23.8
Commercial CPUs	9,288	2,934	31.6
Noncommercial CPUs	1,253	158	12.6
Total	10,541	3,089	29.3

Source: U.S. v. IBM, Plaintiff Exhibit 1962A; absolute revenue and profit dollar data also available from Table 8 and Table 9 in De Lamarter, *Big Blue,* p. 73.

Table 5.7 Revenue and Profitability by Submodel for IBM 2400 Series Mainframe Tape Drives ($ millions)

Model and Submodel	Transfer Rate (Kbps)	Total Revenues	Total Profits	Margin (%)
2401				
1	30	$6.1	$-.0	-.0
2	60	8.2	1.3	18.8
3	90	2.1	.8	37.4
4	60	43.7	6.5	14.9
5	120	51.3	15.3	29.8
6	180	29.6	13.2	44.4
2402				
1	30	8.0	-.3	-.0
2	60	25.2	4.6	18.2
3	90	7.7	2.9	37.9
4	60	63.8	7.7	12.1
5	120	132.3	32.8	24.8
6	180	78.5	32.7	41.7
2403				
1	30	8.3	-.2	-.0
2	60	8.3	.7	8.2
3	90	2.0	.4	21.4
4	60	43.9	7.6	17.4
5	120	61.2	13.9	22.7
6	180	26.2	8.4	32.0

Source: U.S. v. IBM, Plaintiff Exhibit 4216; data also available in De Lamarter, *Big Blue,* p. 80.

to extract consumer surplus lies in the observation that price sensitivity tended to vary with the type of submodel used. Users of high-end submodels tended to be existing customers, part of IBM's established customer base with lower sensitivity to price, while users of low-end entry submodels tended to be more price-sensitive new or newer customers.

The older established customer base had become "locked-in" to IBM's mainframes as a result of their existing investment in application software programs.[6] The only way such a customer could upgrade a mainframe system (to provide greater memory capacity, for instance) was to either buy or rent a higher end IBM submodel or switch to a competitor's system of equivalent power. The cost to the customer of converting software application programs to run on non-IBM systems, as well as the disruption to the customers' business activities this caused, made such switching extremely difficult to justify economically. The result was the diminution of product substitutability, competitive intensity, and therefore price sensitivity. IBM exploited this lock-in induced exaggeration of price insensitivity through the abnormally high pricing of the higher end submodels—the upper limits of which are theoretically defined by the costs to the customers from conversion and business disruption—with the result of boosting revenue and profits beyond that implied in a single price policy.

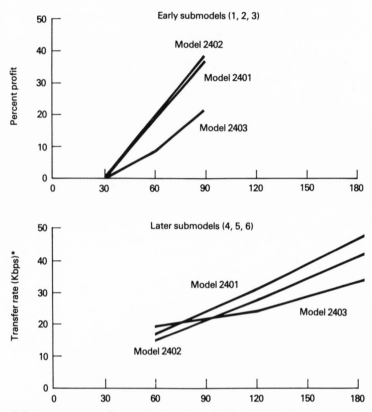

Figure 5.5. Percentage profit margins by submodel for IBM 2400 series mainframe tape drives. *Source: U.S. v. IBM,* Plaintiff Exhibit 4216.

At the same time IBM was able to expand its customer base beyond that implied by a single-price policy by offering low pricing to newer customers. Newer customers, unencumbered by the strategic phenomenon of lock-in and, in particular, the threat of massive conversion-induced losses, had a greater range of feasible purchase choices. In other words, these customers perceived greater product substitutability and therefore were more sensitive to IBM's prices in their purchase decisions. By offering this group prices lower than those implied by a single-price policy, IBM was able to attract more new customers than would otherwise be the case.

Differential pricing disproportionate to cost differences is all that is required to establish the presence of price segmentation. IBM went beyond mere price segmentation, however, especially in its below-cost pricing of the 360 mainframes, with a policy that was not only designed to extract consumer surplus but was also directed at limiting competitive entry and share penetration.

On the surface it seems that such a strategic approach was available to IBM's competitors as well, since lock-in worked both ways. Such lock-in has indeed served to protect the limited existing share of non-IBM compatible mainframes. But, in fact, IBM competitors are unable to copy or effectively respond to IBM's extreme version of segmented pricing at the low end of the market because they do not have the same large base of established customers that provides IBM with the "deep pocket"[7] necessary to sustain such unprofitable pricing at the low end.

In effect, IBM is willing and able to "purchase" customers with the help of the surplus revenue and extranormal profitability extracted from the locked-in customer base. IBM sacrifices these losses on the entry level submodels on the rationale that the long-term loss from acquiring more new customers at below breakeven prices is exceeded by the long-term gain of increasing the size of its locked -in customer base. The net long-term gain is positive because new price-sensitive customers with time, and specifically as their software base grows, become established locked-in customers.

To summarize, the data presented here indicate a pronounced pattern of differential pricing at the submodel level. Further, such pricing behavior is not based on cost differences, but rather is based on a strategy of expanding the customer base and limiting competitive penetration in the most price-sensitive and competitive part of the market while extracting the greatest revenue possible from the less price-sensitive, less competitive, locked-in part of the market.

Because such segmented pricing existed within models at the submodel level rather than at the model level, profit rates on the model level revealed no clear pricing pattern and masked the low price (and below-cost price in the case of the 360) of entry products at the submodel level. Thus an examination of profit rates at the model level (top five lines of previous Table 5.6) revealed no clear price-segmentation pattern across models and camouflaged the price segmentation across submodels.

PRICE SEGMENTATION THROUGH EDUCATIONAL DISCOUNTING AND THE CONTRACTUAL ISOLATION OF MARKETS

Unlike "functional pricing," which does not adhere to the strictest definition of price segmentation in that it entails some product variability, more "pure" forms of price segmentation are discernible in IBM's behavior. Tie-in sales, metering, and bundling (discussed later in this chapter), for instance, can extract consumer surplus without employing some product dimension as a basis for varying price and indeed can extract consumer surplus without explicitly varying price at all. But perhaps the purest form of price segmentation used by IBM is its traditional discount policy toward educational institutions.

In this practice IBM explicitly charges lower prices to educational institutions than to commercial customers. One reason for this is to promote familiarity with IBM systems among students who later may enhance the sale of IBM systems when they enter business. But even without invoking this justification, IBM's two-price policy is entirely rational from an economic standpoint. IBM extracts a higher price from commercial customers who are willing to pay more because of greater income, their aversion to switching brands, and the indispensability of the computer to the continued viability of their business. IBM charges less to educational institutions, which are less inclined to pay IBM premiums due to lower income, a greater willingness to experiment with different brands, and the relatively higher dispensability of a computer needed for educational purposes (as opposed to a business, which buys computers for operational purposes only).

This behavior is an example of pure price discrimination because the products are identical and the differential pricing is explicit. The key to effectively implementing this discriminatory-pricing strategy and assuring extranormal revenue levels lies in the ability of IBM to isolate these two customers groups. If arbitrage existed between these sectors, the Pigouvian assumption of nontransferability between markets would be violated. IBM readily eliminates arbitrage, however, by specifying in the contractual conditions of sale or rental that "noneducational" use is prohibited or, alternatively, noneducational use requires additional payment to IBM. In the conditions of contract for the "educational allowance," IBM specifies that "the educational allowance granted is to assist the educational institution in . . . its internal affairs."[8] There is no extra charge if the product is used for "internal purposes," which are defined to include "use by the faculty, staff, students . . . of the educational institution in instruction and academic research."[9]

Extra payment is required for "external use." Use for "external purposes" is defined to include "commercial research, . . . sale of block time or any work by other than faculty, staff, students or employees of the educational institution." The additional payment for external use is sufficiently high to make arbitrage unprofitable. Moreover, because educational institutions are nonprofit, there is less temptation to violate the conditions of contract by either charging for outside use by commercial firms, or in the latter case, by not reporting such outside use.

Clearly identifiable cases of price discrimination such as IBM's behavior discussed earlier are relatively infrequent because legislation prohibits most cases of discriminatory pricing among customers. However, by presenting this marketing practice as a discount to educational institutions who might otherwise be unable to afford computers and, by implication, some form of charity to nonprofit organizations, rather than "price discrimination" between markets, legal prohibitions have been circumvented.

PRICE SEGMENTATION THROUGH TIE-IN SALES

As Chapter 2 showed, the extraction of consumer surplus can be achieved through behavior that, unlike IBM's "functional pricing" or educational-discount policy, does not involve the use of explicitly differential prices. One of the earliest examples of this nonexplicit segmented pricing in the mainframe industry is IBM's policy of tying the sale of keypunch data cards to an IBM mainframe through the contractual rental arrangement. IBM required that only IBM cards could be used with IBM computers. Violation of this stipulation resulted in the immediate termination of the rental agreement.

This policy was challenged twice by the U.S. Justice Department but, ironically, for the wrong reason. Consistent with the conventional generic legal thinking of tied sales (discussed in Chapter 2), the court argued that such a policy would allow IBM to secure a monopoly position in the card market. But, consistent with the results of business research into other markets, IBM's share of the card market remained well below half and therefore could not be construed as a monopoly.

In fact, IBM's behavior was directed at the mainframe market, not the card market. The card market was the means to an end not the end itself. In particular, the card policy was a vehicle of price segmentation in the mainframe market as opposed to a way of securing monopoly position in the card market. More specifically, the cards served as a device for metering the intensity of usage among mainframe customers and as a device for charging differential prices in the mainframe market. Because, as Chapter 2 explained, use-intensity can serve as a proxy variable for the utility or value a customer ascribes to a product and hence his reservation price, IBM's tying the sale of cards to mainframes was a means of extracting consumer surplus in that the more a customer used a mainframe (and thus the higher the utility associated with the mainframe), the more he had to pay in the form of card purchases. Consistent with this interpretation Sharpe found that "requirements for cards [use-intensity] are . . . well correlated with the value received from equipment."[10]

Theoretically the degree of discriminatory pricing through tie-in sales is positively related to the size of the tied good's (the card's) margins. If cards were sold at cost, no extra profitability could be achieved because the price just covers cost. The greater the card's price over cost, the greater the effective extraction of consumer surplus. By tying the products, the price of the combined date processing system becomes, in effect, the base price of the machine plus the price of all cards used. The greater the card margins, the greater the effect machine-use-intensity has on the total system price. This interpretation is consistent with Sharpe's further finding of extremely high margins of IBM's data cards as well as the finding of Belden and Belden that, during the twenty-year period relevant to our discussion of tied card sales by IBM, "25 percent of IBM profits are attributable to card sales."[11]

IBM's segmented pricing program is therefore, in this case, implemented not through the direct charging of different machine prices, but indirectly through customers' variable use of premium-priced cards. The subtlety of such segmented pricing rests in the fact that IBM charged uniform prices for the machines as well as the cards. Neither market separately could be linked even remotely with price discrimination. But when the markets are considered together it becomes clear that by tying cards to the machines, price segmentation is facilitated despite the ostensibly uniform nature of the pricing behavior.

Nevertheless, on the rationale that IBM was attempting to create a monopoly in the card market, the U.S. Justice Department prohibited IBM from requiring that only IBM cards be used with its mainframes.[12] This "First Consent Decree," however, was ineffective because it permitted IBM the right to require that cards used on its machines meet certain minimum specifications that, conveniently, only IBM cards could meet. According to IBM the requirement was necessary to prevent damage to its mainframes. But, in fact, IBM held patents on the manufacturing equipment that produced over 80 percent of the cards, and the specifications subsequently established by IBM were such that no rival firms could produce cards acceptable for use on IBM mainframes, even though customers were legally permitted to use non-IBM cards. As a result, the First Decree was completely ineffective in achieving its purpose of limiting IBM's power in the card market, although it did have the unintended effect of removing one vehicle to price segmentation: legal contracts tying the sale of cards to mainframes. This source of tied selling have been removed, however, IBM turned to another: specification requirements for cards only IBM could meet.

Eventually, recognizing the ineffectiveness of the First Consent Decree, the Justice Department brought suit against IBM again (over twenty years after the first suit was brought). The issue was again identified as monopoly power in the card market rather than price discrimination in the mainframe processor market. High margins were identified on IBM cards and were used as evidence for monopoly power in the card market rather than a tool for extracting surplus revenue in the CPU market.

Among other provisions of the Second Decree, IBM was required to prove that any differentials in the price of its cards were based only on "differences in the cost of manufacture, sale or delivery." The objective of the provision was to eliminate "monopoly" pricing in the card market by eliminating IBM's proprietary card-production-process advantage. In particular, IBM was required to sell up to thirty of its rotary presses each year for five years at reasonable terms.[13] And if the company could not convince the court that "substantial competitive conditions existed" in the card market, IBM was to divest itself of any production capacity in excess of 50 percent of the total U.S. capacity.[14] Moreover, IBM was required to offer its patents under nonexclusive, unrestricted license to any applicant at a "reasonable" royalty.

Hence, while the First Consent Decree precluded the direct tying of card sales to mainframes, but tolerated indirect tying facilitated through IBM's card specification requirements and its proprietary card production process, the second decree not only eliminated the high premium on IBM cards by requiring that premiums be related to manufacturing cost differences but also eliminated IBM's proprietary card production process. The unintended net result of the decrees was to effectively eliminate the possibility of using cards as a vehicle of price segmentation in the equipment market, although the intent of the decrees focused exclusively on reducing monopoly power in the card market.

The complex irony of this whole analysis is that the provision of the Second Decree did achieve one intent of the antitrust legislation: nondiscriminatory pricing. But this effect was achieved accidentally with a decree designed to achieve a different immediate end, elimination of monopoly position in the card market. The further irony is that this unintended effect of removing discriminatory pricing through tie-in sales did not reduce IBM's market share in the *equipment* market. Even if all forms of price discrimination were removed from the mainframe processor market, monopoly power would not necessarily be reduced because price discrimination is a symptom not a source of "monopoly power." The final irony is that even though the government efforts prevented IBM from price segmenting through tied sales (albeit accidentally), price segmentation, in fact, retained a significant role in IBM's behavior in the processor market, for IBM found another means to this end: extra-use charges.

PRICE SEGMENTATION THROUGH METERING:
EXTRA-USE PRICING ON THE IBM 709 SERIES THROUGH
THE IBM 1130 SERIES

Immediately after the Second Consent Decree became final, IBM put into place a rental scheme that involved the charging of customers extra use fees outside a designated "prime-shift" use period. Specifically, IBM charged a base monthly rental fee for use of a computer inside the prime shift period. Outside the prime shift, customers were charged by the hour for every hour in an amount equal to 40 percent of 1/176 of the monthly fee. (1/176 is presumably used because the denominator represents the hours in a prime shift of a rental month).

Before considering the evidence supporting the position of this study that this practice represents a form of price segmentation, two alternative interpretations should be noted. First, since maintenance was included in the monthly rental fee it can be argued that extra-use charges merely reflect the higher maintenance associated with higher computer utilization. Second, and relatedly, it can be argued that the extra-use charges merely reflect the more rapid depreciation of more intensely used machines. The relevant question for the presence of price segmentation behavior is not whether or not these other sources of higher costs for more intensely used computers exist, but whether the amount of these costs fully accounts for the extra-use charges and thus the higher prices paid by more intense users. If the higher amount paid by high-utilization customers is not fully explainable in terms of cost differences then the behavior is a form of price segmentation.

The first alternative position, that extra-use charges only reflect higher maintenance, can be readily tested and resolved by removing the maintenance price to the customer from the customer's total price and observing if price differentials remain. The related position, that extra-use charges represent depreciation for more intensely used machines, cannot be incontrovertibly disproved for lack of detailed, actual depreciation cost information.

Two points, however, should be noted concerning this second position. First, to the extent that maintenance is effective, *actual* depreciation is reduced, if not eliminated. In fact, for the last twenty years it has been the explicit position and policy of IBM that the "periodic replacement of assemblies with new components, either as part of a preventive maintenance plan or upon failure, can serve to keep a system almost as good as new."[15] Moreover, the accounting methodology used by IBM for calculating the depreciation of equipment subject to extra-use charges treats depreciation as a simple time function rather than a function of variable machine-usage rates. It would thus be inconsistent in terms of IBM's costing procedures and accounting methods to justify higher charges for higher utilized machines on the basis of greater depreciation. Finally, it should also be noted that a third possible position that explains extra-use charges as the combination of higher maintenance costs and more rapid depreciation is logically flawed to the extent that it views maintenance and actual depreciation as additive. In fact, greater maintenance is, within limit, logically associated with reduced actual depreciation.

Data taken from IBM General Service Agreement Schedules reveals a price differential between high- and low-use customers across a wide range of mainframe systems. The great magnitude of the differentials suggests that price differentials more than reflect possible cost differentials despite the inability to account for actual depreciation. Figure 5.6 gives the ratio of the price charged high-utilization customers to the price charged low-utilization customers. High-utilization customers are logically but arbitrarily defined as those who operate the system for three shifts, while low-utilization customers use the system during prime shift only. The ratios actually represent a ratio of rental figures, each consisting of a base rent only; maintenance cost is removed. As can be seen, in most cases, the prices charged to more intense users of the CPUs are generally at least 60 percent, and typically 70 to 80 percent, higher than those prices charged to less intense users.

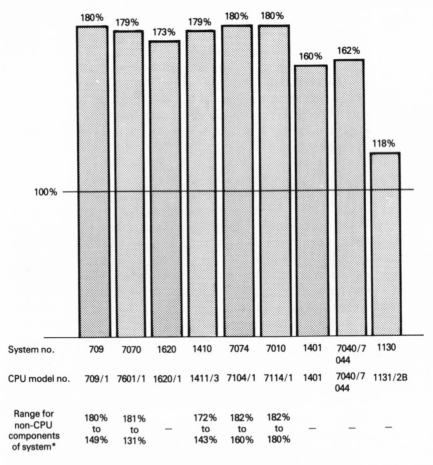

System no.	709	7070	1620	1410	7074	7010	1401	7040/7 044	1130
CPU model no.	709/1	7601/1	1620/1	1411/3	7104/1	7114/1	1401	7040/7 044	1131/2B

Range for non-CPU components of system*	180% to 149%	181% to 131%	—	172% to 143%	182% to 160%	182% to 180%	—	—	—

*Includes tape drive, disk drive, tape controller, memory, terminal, printer, and other mainframe system components

Figure 5.6. Ratio of charges to high-use IBM customers versus low-use IBM customers for IBM CPUs and mainframe peripherals. *Source:* Calculated from IBM General Service Agreement Schedules.

charging for software usage and of relating software prices to the price of hardware used both facilitate price segmentation in that software usage or hardware price is a proxy for the value the customer ascribes to the software product. Because virtually all of the cost of software is fixed development cost[16] and thus marginal cost is virtually nil, price segmentation is especially attractive in the software market because all surplus revenue extracted, *including that from additional unit sales to the more price-sensitive customers,* translates directly into profits.

PRICE SEGMENTATION THROUGH "STABLE CHARGES": PRICING THE IBM 370 MODEL 15

The history of pricing activity by IBM in the mainframe market up to and including the 370 mainframe product generation is marked by a rigid policy of not changing the price of a particular generation from its introduction through its maturity. IBM labels this practice its "stable-charges" policy. Sharpe offers two possible explanations for it:

> Increased reliability and software support may make older equipment as effective overall as newly designed models of comparable cost. IBM may be particularly wary of Justice Department disapproval of price cuts, preferring to eventually scrap its used systems rather than to reduce prices to keep the equipment competitive.[17]

In addition to the possible increased effectiveness of older software and wariness of antitrust violations, this thesis suggests another reason for the stable-price policy: it facilitates price segmentation. The key to understanding how a stable charges practice can facilitate price segmentation lies in the observation that the simultaneous market presence of two product generations at different price/performance levels is in fact differential pricing with respect to computing power. In a purely price-competitive market, old equipment prices will drop as new generation technology that offers superior price performance is introduced. By holding prices constant through the life of a given product, after a new product generation of superior price/performance is introduced, a tiered-pricing program is implemented, which to the extent that consumers continue to buy or rent the older generation product, extracts consumer surplus. Regardless of the customer's reasons for retaining the older product—whether it be abnormally high levels of risk aversion to new technology, perceived conversion costs and disruption to the customer's business, or simply satisfaction and thus complacency with the current system—the willingness of some customers to pay high prices relative to computer performance means that pricing that exploits this form of customer heterogeneity will result in greater revenue than a policy of making the price/performance of old products consistent with the new.

Testimony and confidential IBM documents disclosed at the trial hearings for the law suit brought against IBM by Telex Corporation reveal that IBM's stable charges policy did, in fact, result in a tiered pricing structure that had the effect of charging and extracting more from those customers willing to pay more.[18] This comes to light as a result of the particular disclosure of IBM's internal strategic analysis concerning what type of pricing strategy should be used for the IBM System 370 Model 15. In this analysis IBM confirmed the effectiveness of a "stable charges" policy for two sets of overlapping product generations.

Concerning the first set of product generations, IBM determined that a large base of IBM customers were paying substantially more for pre-360 mainframes than 360s of comparable power because IBM did not reduce prices on the earlier, pre-360 equipment, even after 360 systems with superior price/performance had been widely installed. The overlap of the 360 and 370 generations followed the same pattern. The 360 price was held constant with the result that three years before the 370 was introduced, the 360 price began to exceed the market price. By the first year of the 370 introduction (seven years after the 360 was introduced), IBM determined that the 360 was substantially below market price/performance levels. To compensate for the risk of share loss on the 360, the 370 was priced substantially below market price/performance levels. This allowed IBM to extract higher prices from the customers unwilling to upgrade from the 360, while simultaneously expanding the customer base through the 370's appeal to more price-sensitive customers.

Despite its initial below-market level, the 370 price was set sufficiently high so that, according to IBM's price and performance forecasts, it would intersect the declining market price three years before the introduction of the follow-on product to the 370, the "FS." This three-year intersection point was carefully chosen considering such factors as the extent to which the announcement of the planned introduction of the FS would prevent customers from switching to superior price/performing competitors in anticipation of IBM's new system. Balancing the financial impact of the risk of losses at the end of the 370 cycle before the follow-on was to be introduced against the gain from the tiered price structure, IBM concluded that a stable price policy with the recommended intersection point resulted in 74 percent higher revenue than a declining market-price policy. In this way a series of discrete, horizontal (unsloping) price–performance lines, which jump from above to below the market price level between overlapping product generations, facilitated the extraction of excess revenue from those willing to pay it and the expansion of sales among price-sensitive customers without any explicitly differential pricing.

A stable price policy is also used to exploit differences in customer's price sensitivity for older equipment arising from geopolitical factors or customer sophistication differences. The lack of experienced data processing personnel able to take advantage of the new potentialities of the latest computer technology in, for instance, developing economies can artificially inflate prices for older generation equipment. Similarly, restrictions on computer sales to Soviet-bloc nations and other countries, which effectively remove the newest technology from the marketplace in these countries, also create a willingness to pay higher prices for older equipment than prices in markets where the older equipment must compete with the new. By combining a stable price policy with marketing efforts focused on selling returned old generation equipment to less-developed nations and nations restricted from new technology sales, IBM is able to exploit the willingness of such customers to pay higher prices for inferior price/performance machines without losing more price-sensitive customers in more competitive markets.

PRICE SEGMENTATION THROUGH BUNDLING

Product bundling—the practice of offering two or more products for a single price—manifests itself in various forms in IBM's behavior. IBM bundles hardware with software, peripherals with processors, peripherals with other peripherals, as well as support and service with any of its equipment. Some of its most strategically significant bundling activity has included specifically bundling CPUs with operating systems, bundling CPUs with "free" application software packages, bundling CPUs with controllers, bundling disk drives with controllers, bundling disk drives with data storage modules, and bundling all this hardware with many and varied types of customer support.

While past legal decisions and the threat of future public and private antitrust suits have forced upon IBM a cautiousness in its bundling activity, IBM documents disclosed in trial reveal that IBM aims "to be the most bundled unbundled vendor possible within legal constraints."[19] Such bundling is desired because, in addition to creating structural constraints inhibiting competitive entry and mobility, it also serves, in various ways and without explicit differential pricing, as a means to price segmentation. This next section shows the applicability of the formal generic analysis of bundling (discussed in Chapter 2) to the mainframe market; the section following reveals another way that bundling facilitates price segmentation, not addressed in the formal generic analysis of bundling.

Exploiting "Preference Reversal": Bundling IBM 370 Models 158 and 168 with Memory

While the presence of bundling behavior in the computer industry is clearly observable, the various means by which bundling facilitates price segmentations are not widely recognized. Perhaps the most subtle of these means lies in the principle of "preference reversal" first identified by Stigler. As discussed in Chapter 2, bundling can facilitate price segmentation not by simply exploiting absolute differences in customer price sensitivities for one product but by exploiting relative differences in sets of price sensitivities for two or more products in a bundle.

Figure 5.9 lists the prices IBM customers are willing to pay for the individual components of an IBM bundled product: the IBM 370 Model 168. As indicated, the processor is more valuable to both firms than the memory. But also notice that AT&T places a higher value on the 370 processor than does GE, while GE places a higher value on the one megabyte of memory than does AT&T.

To sell both items to both firms without bundling, IBM could set prices no higher than $40,500 for the processor and $7,500 for the memory. The total revenue in this case would be $96,000. IBM actually charges one fee, $53,800, for both. AT&T and GE still buy the bundle because $53,800 does not exceed the total either are

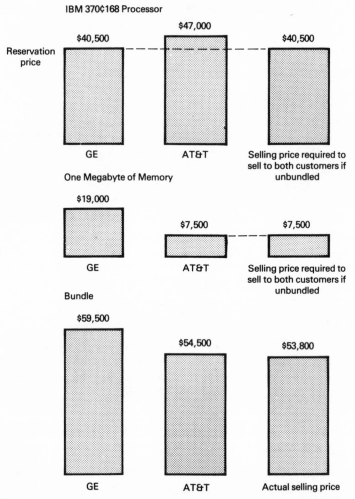

Figure 5.9. Customer reservation prices and actual selling price of the IBM 370 Model 168. *Source:* IBM customers; customer names have been disguised to protect proprietary data.

willing to pay for the bundle. Yet IBM's total revenue in this case ($107,600) exceeds the revenue derived from an unbundled price arrangement ($96,000).

This activity is a form of price segmentation because it involves the extraction of the consumer surplus by effectively charging the customer firms different prices for the same product. In particular, customers are effectively charged different prices for the memory and different prices for the processor. This is illustrated in Figure 5.10. The revenue received by IBM under an unbundled pricing strategy is the $96,000 represented by the dark grey area. Bundling allows IBM to effectively receive a higher price from AT&T than GE for the CPU. Furthermore, bundling allows IBM to receive more from GE than AT&T for the memory. The result is the extraction of consumer surplus of, in total, $11,600 (the total striped area).

Note that the precise level of extraction from each customer separately is indeterminate; the diagram could have been drawn with AT&T retaining some level of its consumers surplus and thus GE retaining less, but the aggregate effect on IBM—a gain of $11,600—remains the same irrespective of how the loss to the customers is distributed. Also note that if the *total* customer reservation price for each customer were the same (even though the preferences for the components of the bundle differed), IBM could achieve first-degree price discrimination by charging actual prices equal to reservation prices. Different total valuations imply at most second-degree consumer surplus extraction levels.

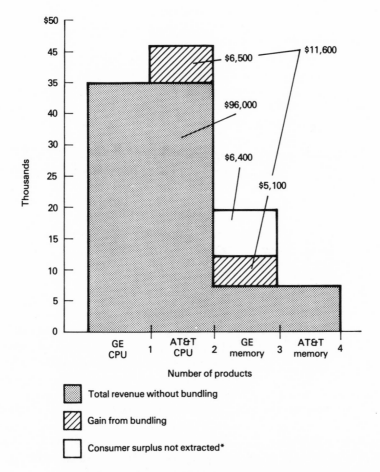

Figure 5.10. Revenue gain from bundling the IBM 370 model 168. *Source:* Calculated from data in Figure 5.9.

Price Segmentation through Bundle Variability:
The Concentration of Customer Support

As Chapter 4 discussed, a prominent feature of IBM's behavior is the bundling of hardware with various types of customer service and support. IBM has argued this is motivated by the efficiency consideration of billing simplicity. "These services could have been priced separately," explains the former head of the IBM pricing group, but "we chose to use the box [computer] as our billing vehicle for the total package, thus establishing the box price as our sole means of recovering our investment and return on that investment." This was more "convenient" than discretely pricing the hardware and each of the various types of support.[20]

The bundling of customer support into the price of IBM hardware serves as a vehicle to price segmentation in that different customer groups receive different levels of field support for the same price. On the surface this appears to be in inversion of the basic price discrimination concept of offering the same product at different prices. But in charging the same price for different levels of support, a seller is also charging different prices per unit of support. Given this tautology, empirical evidence that (in the absence of price differentials) customer support varies by customer segment points to the presence of price segmentation and the associated extraction of excess revenue.

Data on the IBM field force and customer base for the IBM 360 series mainframe processor recently made publicly available as a result of disclosures in *U.S. v. IBM* provide convincing evidence of significant variances in field support. Variation in IBM field support is revealed by comparing the revenue value of IBM's business contributed by different customer segments with the level of IBM field support received by these different segments. Field-support variances are measurable in terms of man-years (the total number of years a given number of men work), and customer segments are defined by account size (the amount or revenue they contribute to IBM). The amount of support received relative to the amount of revenue paid varies dramatically by customer segment.

IBM's customers are characterized by great variability in the amount of business they contribute to IBM. In particular for the 360 series, as the left and middle bars in Figure 5.11 reveal, only 3% of IBM's customers account for over 60% of IBM's business and only 1.5% of its customers account for over 50% of its business. IBM's business is thus highly concentrated among only a very small number of very large customers. Conversely almost 90% of its smallest customers account for less than 15% of its business and over 70% of its customers account for less than 5% of its business. Through bundling IBM is able to exploit this especially heterogeneous feature of the customer base.

Bundling permits IBM to vary support so as to favor the smallest customers and provide diminished support levels to the small number of very large customers. Comparing the middle and right bars in Figure 5.11 shows that the group of largest customers (dark grey) that contribute over *half* of IBM's revenue, receive only about *one-third* of all man-years of field support. Similarly, while the top two largest customer segments (dark grey and adjacent striped area) contribute over *three-fifths* of IBM's business, they receive less than *half* of all field support. Conversely, the

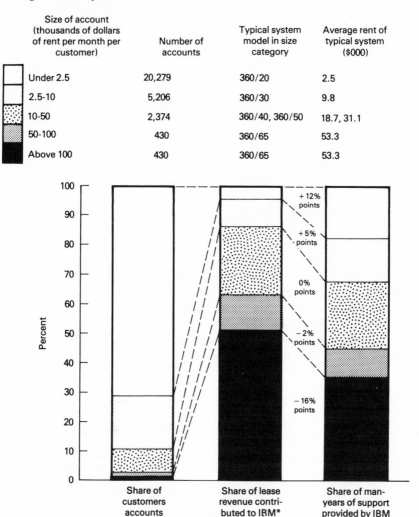

Size of account (thousands of dollars of rent per month per customer)	Number of accounts	Typical system model in size category	Average rent of typical system ($000)
Under 2.5	20,279	360/20	2.5
2.5-10	5,206	360/30	9.8
10-50	2,374	360/40, 360/50	18.7, 31.1
50-100	430	360/65	53.3
Above 100	430	360/65	53.3

*Percentage of IBM's total monthly rental revenue that IBM collected from each size category.

Figure 5.11. Distribution of revenue contributed and support received by IBM's customer base. *Source:* Calculated from *U.S. v. IBM*, Plaintiff Exhibit 5310A.

smallest size customers (light grey) who comprise less than *5 percent* of IBM's business receive almost *one-fifth* of all field support man-years. Similarly, the two smallest customer groups (light grey and adjacent striped area), which account for less than *15 percent* of IBM revenue receive about *one-third* of all the field support. The middle-size customer segment receives IBM support directly proportional to its contribution to IBM revenue.

The percentage point differences between what the customer segments pay in revenue and what they receive in man-years is listed between the middle and right bars. The twelve percentage point difference between what the group of smallest customers pay and what they receive is very substantial. For instance if the customer group were to pay IBM an amount proportional to what it receives, it would have to contribute an additional $1.1 thousand million in revenue to IBM, an amount equivalent to over a third of the total profits received by IBM (12.1% of $9.3 thousand million in total revenue). Conversely the largest customers pay 51% of IBM total revenue—16 percentage points over the amount indicated by their share of field support. This differential means they overpay by $1.5 thousand million in revenue (15.8% of $9.3 thousand million), an amount equivalent to over one-half the total profits received by IBM. The amount of excess revenue contributed by large customers, and the amount small customers do not contribute to IBM as a result of their preferential treatment, is diagrammed in Figure 5.12.

Comparing revenue paid per man-year received in each segment to the total average rate across all segments illustrates how varying field support through bundling is a vehicle to price segmentation. Figure 5.13 shows that the smallest customers pay less than one-third the rate for field support than all customers, on average, pay while the largest customers pay almost 50 percent more than the average rate among all customers and almost five times the rate the smallest customers pay. The gradation of rates is strikingly even, with the customers in the middle-size category of the five categories paying exactly the same rate that all customers, on average, pay. (The actual rates are listed below the bars.)

Why is IBM able to provide less support to large customers than small customers for the same price and thus effectively charge more per unit of support to large

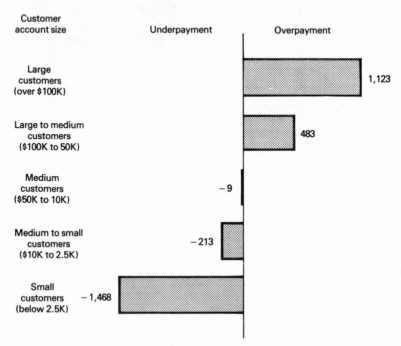

Figure 5.12. Revenue paid in excess of share of field support received ($ millions). *Source:* Calculated from *U.S. v. IBM*, Plaintiffs Exhibits 1962A, 5310A.

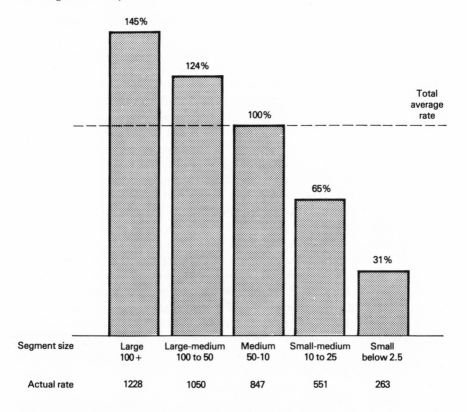

*Rate is dollar revenue paid to IBM per man-year of IBM field support received.

Figure 5.13. Revenue paid per man-year received: segment rate relative to total average rate. (Rate is dollar revenue paid to IBM per man-year of IBM field support received.) *Source:* Calculated from *U.S. v. IBM*, Plaintiff Exhibit 5310A. Note that the relativity of segment rates to the total average rate is directly obtained (without requiring actual revenue per man-year rates) by dividing revenue percentage share by man-year percentage share for each segment. Actual rates need not be used because they are implicit in the share percentages. Let SR denote segment revenue, TR total revenue, SM segment man-years, and TM total man-years:

Revenue % share/man-hour % share
= (SR/TR)/(SM/TM)
= (SR/SM)/(TR/TM)
= Segment rate/total average rate.

customers? The simple answer is large customers tend to be more self-sufficient. More specifically, greater in-house data processing expertise means the consequences of product failure in the absence of external support are less severe. This results in lower perceived risk and a relatively lower perceived need for IBM's support.

Large customers tend to have internal support staff (often consisting of a large number of former IBM field personnel). Because a certain level of data processing activity is required before a staff can be kept busy full time, the larger a customer's data processing operation, the more likely it is able to cost justify a central support staff and hence the more self-sufficient it tends to be. Moreover, the greater the scope of the operations, the more likely more and more specialized data processing personnel can be retained on a viable economic basis. To the largest companies, such as AT&T, IBM field personnel are not only redundant but positively detrimental to efficiency. The IBM support staff dedicated to AT&T thus consists largely of IBM account managers as opposed to IBM systems-support personnel.

The criticism has been raised by IBM lawyers that this interpretation of IBM's behavior as price discrimination is contrary to the economic logic of price discrimination and, specifically, the logic of charging more to those willing to pay more. They suggest that varying support levels in favor of small customers implies that the largest customers who need support the *least* are willing to pay the *most* for that support—an inversion of price-segmentation logic in which those *least* willing to pay are charged the *least*.

Regardless of whether such behavior is price discrimination from a legal standpoint, it is, in fact, a sophisticated form of price segmentation in the economic sense. The validity of this assertion lies in understanding the unit of analysis. The unit of analysis, in this case, is not the field support separately but the total product bundle of hardware and support combined. Large customers are more willing than small customers to pay a higher rate, not for support, but for the total bundle. IBM is not charging large customers more for support services but is charging large customers the same amount as small customers for an inferior bundle of goods. In effect, therefore, IBM is charging more to large customers for an *equivalent* bundle of goods than it charges to small customers.

Even though large customers value support less, they value the total bundle more because of the higher value they place on hardware. In other words the suggestion here is that the larger customers' lower price sensitivity for hardware overwhelms its higher price sensitivity for support, with the net effect that the larger customer is less price sensitive than the smaller customer for equivalent bundles.

Low hardware-price sensitivity makes sense for larger established customers for several reasons: (1) they are already locked into IBM hardware as a result of committing substantial resources to applications; (2) mainframe processors are indispensable to large customer operations (unlike small customers in some cases); and (3) the absolute level of profits (not margins) are greater, implying a less severe income constraint. In the absence of empirically measured elasticities for hardware, customer support, and bundle equivalents (by customer segment), the notion that large customers tend to be less sensitive to the overall bundle than small customers seems reasonable and does fit the facts: IBM provides less to large customers for the same price and, in effect, charges more to large customers per equivalent unit of the product bundle.

In an economic sense, this is price segmentation; in a legal sense, it may not be. While this study in no way attempts to resolve the legal debate it should be noted that expert witnesses for the defense in *U.S. v. IBM* argued, in effect, that although IBM's behavior resulted in varied levels of support after the purchase of the bundle was made (and therefore constitutes price segmentation as an *ex post* economic phenomenon), it does not constitute "intended" price discrimination. Fisher, McGowen, and Greenwood explain their testimony:

Variation in support is not in fact price discrimination. . . . The reason for this is clear if one understands the difference between risks *ex ante* (before the outcome of a risky event is known) and risks viewed *ex post* (afterwards). In providing the bundle, manufacturers accept risks concerning the proper installation and use of their products. In effect, they are providing a warranty for customers. Not very surprisingly after the event, it may turn out that some customers appear to benefit more from that warranty than do others. That is the nature of the acceptance of responsibility for the outcome of uncertain events. After those events become known and are no longer uncertain, some of them may turn out to be favorable and others unfavorable. That some customers turned out not to need the bundled software and support as much as others is only a reflection of this fact.[21]

This argument is noted here, not to represent this study's support for any legal position, but to illustrate that the evidence presented here for price segmentation, in the economic sense, does not necessarily imply a parallel legal position.

Bundling allows IBM to vary support in such a way as to effectively charge a lower price to small accounts and a higher price to large accounts without explicitly charging different prices. The price differentials are hidden within the constant single price of the bundle. If the support were priced separately, IBM could certainly provide varying levels of support, but differential support levels would have to be reflected in differences in separate support charges to the customers.

Prices to smaller customers requiring high levels of support would thus soar. This in turn would, at the least, open a wide door to competitors and, worse, could completely eliminate a large number of customers from IBM's reach simply because the customers could not afford the higher prices. Bundling allows IBM to reach these small customers with lower effective prices without explicitly charging differentially lower prices.

Apportionment Methodology for Revenue Costs

The mask of bundling that IBM used to disguise segmented pricing is complemented and reinforced by the camouflage of IBM accounting methods. IBM's accounting methodology, disclosed in the U.S. Justice Department case, suggests that all customers received an amount of field personnel support directly proportional to the amount of revenue each paid to IBM. IBM distinguishes three types of overhead costs: engineering costs, product costs, and revenue costs. The issue here is the treatment of revenue costs: the overhead cost of the field support associated with the generation of revenue.

According to IBM's Grey Books, revenue costs are apportioned at a constant "normal" rate of 28.69 percent (irrespective of customer size) for IBM 360 models 20, 30, 40, and 50. This constant apportionment rate implies the cost to IBM of field support for each dollar of revenue generated is the same among all customers, large or small. Since IBM charges the same price to all customers for the hardware/support bundle, any other accounting methodology would imply price segmentation; different apportionment rates imply variation in support levels. If, for instance, IBM allocated overhead revenue costs in proportion to man-years of field support received, it would be admitting to suspiciously low, if not negative, margins among its small customers, while revealing enormous profits among its largest customers.

It might be argued that a constant apportionment rate is used as a matter of simplicity in calculation. But, in fact, IBM does not use a constant rate; it uses different revenue apportionment rates. These rate differences, however, are not used to account for variation of support among customers; rather, different rates are employed primarily "due to internals" (listed in Table 5.8). "Internals" refers to the very substantial number of larger model 360 processors IBM itself uses for internal accounting, other administration, and product testing. Since these IBM-operated CPUs required no external support from the IBM field force, IBM reduces apportionment rates accordingly.

When customers are considered separately from internals, however, it is clear that the IBM accounting methodology implies that support levels do not vary across customers (top line, Table 5.8). This constant rate not only implies constant support levels but also allows IBM to price its products as if this were the case. In particular, IBM can set prices at below actual costs to the price-sensitive customers receiving disproportionally large amounts of support, although (using IBM's constant apportionment rate) prices appear to be set at profitable levels.

The skewed profitability across customer segments remains hidden in aggregate data that IBM provided to the courts to show model by model profitability (Table 5.6). In particular, because revenue and profits refer to the total bundled price and because IBM accounting methodology does not account for variability in support levels, the tendency for larger models (primarily used by larger customers) to exceed the profitability of smaller models (primarily used by small customers) remains veiled. Profit data based on constant-rate revenue-cost apportionment and bundled prices therefore do not reveal the strong price-segmentation pattern to IBM's behavior.

Table 5.8 Apportionment Rates for Overhead Revenue Costs

Model	20	30	40	50
"Normal" apportionment rate (without internals)	28.69%	28.69%	28.69%	28.69%
Adjustment "due to internals"	0%	1.09%	3.13%	10.69%
New apportionment rate (with internals)	28.69%	28.38%	27.82%	25.92%
Unexplained adjustments	1.69%	1.38%	0.82%	2.92%
Official IBM apportionment rate	27.00%	27.00%	27.00%	23.00%

Source: IBM Grey Books and *U.S. v. IBM*, Plaintiff Exhibits 1962A, pp. 532, 535, 537.

This pattern is similar to that revealed by observation of IBM's differential sub-model pricing, whereby IBM essentially "buys" new customers in the more competitive part of the market with the profits from large, established, locked-in customers in the less competitive part of the market at prices competitors are unable to sustain. Note that the tendency for smaller customers to be new customers is reflected by the facts that the bulk of IBM field-force support focuses on small customers (Figure 5.11) and that the composition of IBM's field force is heavily skewed toward field employees specializing in systems installation (most useful to newer customers) as opposed to, for instance, system maintenance (Figure 4.3 in Chapter 4).

THE ROLE OF INTERFACE MANIPULATION IN COMPUTER MARKET PRICE SEGMENTATION: ISOLATING CUSTOMER SEGMENTS IN THE IBM 2300 SERIES DIRECT ACCESS STORAGE DEVICE (DASD) MARKET

The extreme plasticity of computer system circuitry permits a wide range of possible locations and formats for connecting peripherals and processors. Modifying the nature of the connecting point between two pieces of hardware and/or changing the "location" of that connecting point within the electronic circuitry of a total system not only creates a barrier to competition, but also allows IBM to effectively isolate its own customer segments from each other, creating the opportunity for differential pricing. This is precisely the type of behavior that occurred in the middle- and low-end disk drive market in the transition period between IBM 360 processors and 370 processors.

The strategic rationale for this particular IBM action lies in the competitive problem that PCMs (plug, or peripheral, compatible manufacturers) posed for IBM at the time. The PCMs were rapidly gaining share through the imitation of newly introduced IBM peripherals coupled with penetration (low) pricing.[22] The share threat was focused on peripherals that attached to new (as opposed to mature) IBM mainframes because the expected life of the market for these peripherals was longer and therefore the return on investment to competitors from imitation was more likely to be sufficient to justify entry.

The market for peripherals that attached to the 360 was thus secure, while the peripherals market for the 370 CPU was likely to face severe share attacks. Given this competitive context, IBM desired to maintain existing high prices and revenue on the disk drive for the large installed base of maturing 360s, the "2314 DASD" (Direct Access Storage Device), and to start a lower but "stable" price line for the 370 CPU disk drives in order to reduce share penetration threats. The problem was that both CPUs used the same disk drive. Frank Carey, IBM Chief Executive Officer (CEO) and Chairman at the time, after a discussion of strategy for this disk drive market, concluded to T. Vincent Learson, the previous IBM CEO, that a "repackaged 2314 is [the] best alternative."[23] Hence, IBM unveiled the 2319A, a repackaged 2314.[24]

The original 2314 was actually a combination disk drive/controller product bundle[25] that connected to the IBM 360 central processor. IBM treated the components of the DASD bundle as separate products internally, and indeed the products remained physically separate until they were shipped to the customer site where they were bolted together. To the external customer market, however, IBM only offered the products together for a single price. This was done primarily to deter customers from substituting competitive disk drives that could attach to IBM's 360 processors.

The 2319A was in fact the 2314 minus part of the controller included in the original bundle. This part of the controller was now mounted in the 370 processor itself and labeled an "in-board file adapter." (See A of Figure 5.14.) IBM changed the disk drive/CPU interface—an interface competitors wrongly assumed to be immutably fixed—by changing the location of the connection. More circuitry was put inside the CPU; the connection now occurred at a level closer to the physical disk drive apparatus than was previously the case.

IBM maintained that this interface change saved costs because the CPU could perform more efficiently the functions previously performed by the controller. This may be true, but the more significant effect was on IBM's competitive and customer posture. The product action cut IBM's DASD customer base sharply into two segments: 360 customers who could use only the higher-priced 2314, and 370 customers who could purchase the functionally identical product, the 2319A, at a lower price. As B of Figure 5.14 shows, new 370 customers paid over 40 percent less to purchase a functionally equivalent disk-storage device and just under 40 percent less to rent a functionally equivalent storage device. To maintain such a large price differential IBM had to make sure that the low-priced 2319A could not be used by 360 customers, otherwise 360 customers would switch to the less expensive 2319A and IBM's disk drive revenue would be reduced potentially by 40 percent. The interface change eliminated any possible intercustomer segment leakage.

By aggressively pricing this repackaged disk drive,[26] IBM made it more difficult for competition to penetrate the emerging disk drive market for the 370 while at the time fully maintaining the higher revenue flow from the existing 360 processor disk drive users. Furthermore, IBM's action shut out existing disk drive competitors in the 360 market from the 370 disk drive market.

Unlike bundling and tied sales, the implementation of which has been performed more cautiously and subtley as a result of past legal decisions, interface manipulation is carried out openly as a result of the dismissal of the recent *U.S. v. IBM* case as "without merit." It remains today a prominent means to segmented pricing. For instance with the recent introduction of the IBM PC System Two Series on 4 April 1987, IBM also introduced a new bus (internal communication standard) that makes it impossible to attach previously compatible circuit boards manufactured by IBM's competition to IBM's computers.[27] This prevents customers from upgrading their computer with competitive add-on circuit boards at a relatively low cost and allows IBM to extract higher margins. In a pattern now familiar, IBM then charges very low prices on its low-end machines, which in this case, reduces competitive threats from low-cost computer "clone" manufacturers in the intensely competitive market for the more price-sensitive customer.[28] The result is greater revenue and profits to IBM by manipulating and exploiting the heterogeneity of customer reservation prices.

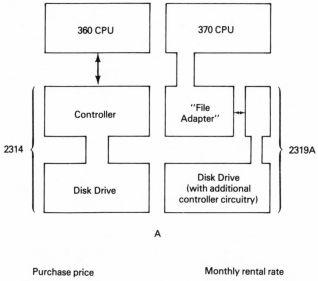

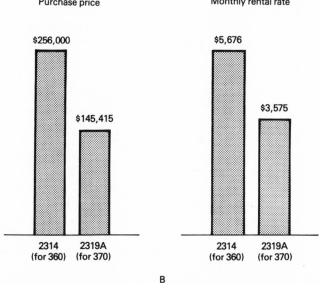

Figure 5.14. System interface differences and price differences for the IBM 2314 and the IBM 2319A mainframe disk drives. *Source: U.S. v. IBM,* Plaintiff Exhibits 6227, 6228.

Interface manipulation can be thought of as a subset of any actions that create data processing system incompatibilities either between products used by different customer segments within IBM's customer base, or between IBM products generally and competitive products. In either case, IBM is not merely exploiting existing differences in customer price sensitivities, but in fact, manipulating and, in particular, creating or amplifying differences in price sensitivity. In other words, to the extent that IBM creates, not merely exploits, the phenomenon of lock-in either internal to its customer base or relative to competition through product actions or contractual agreements, IBM is not simply exploiting customer heterogeneity but is also manufacturing or exaggerating that heterogeneity. Consistent with the theoretical distinction between the two sides of each strategy type discussed in Chapter 3, it is thus both the ability to alter the product's price and the ability to modify the nature of customer price sensitivities that lies behind the effectiveness of IBM's price-segmentation strategy.

CHAPTER SUMMARY

This chapter provided evidence that price segmentation plays an important strategic role in IBM's behavior in the industrial market for mainframe systems, despite legal constraints and in the absence of the measurement of actual elasticities. It showed the applicability of generic analytic frameworks of price segmentation to actual, observed behavior in the mainframe computer industry: tying card sales to mainframes, metering through extra-use charges, and bundling processors with memory. It also revealed other methods of segmented pricing: the "functional pricing" of submodels, educational discounting, "stable charges" across product generations, interface manipulation, and the variation of bundles through the disproportionate allocation of field support.

The presence of a pattern of segmented pricing in IBM's behavior is hard to discern because in all these cases, with the exception of functional pricing and educational discounting, price differences are not explicit. In the case of functional pricing, model-level profit data hide the pattern at the submodel level. The characterization of educational discounting as a charitable action benefiting nonprofit institutions, the idea that maintenance and/or accelerated depreciation fully explain higher extra-use charges, the use of constant-rate-apportionment methodology, and the tying together, bundling, or repackaging of products all serve to camouflage the presence of segmented pricing. Irrespective of the legality of IBM's behavior, this chapter shows that such pricing is not based on cost but rather is based on the strategic objective of protecting or expanding sales among more price-sensitive customers and extracting maximum revenue from less price-sensitive customers.

NOTES

1. *U.S. v. IBM*, PX 2508A; memo dated November, 1969.
2. *U.S. v. IBM*, PX 1001A, p. 14.
3. Chapter 6 develops this point further with specific reference to supercomputer customers and IBM's 360/90 supercomputer not included in the above data. Note, however, that the noncommercial market, which primarily consists of customers with moderately difficult scientific or engineering data processing problems that require different types of features, functions, and performance, should not be confused with the "supercomputer" market niche, which consists of a very small segment of customers within the broader noncommercial market with an extraordinary demand for the fastest possible computing speed.
4. *U.S. v. IBM*, Northrop, the IBM executive responsible for pricing methodology, Tr. 82232, 82247.
5. *U.S. v. IBM*, PX 2508A.

6. See Geruson, "Elements of Strategy in IBM's Price and Product Behavior," for a review of the dynamics of competitive rivalry associated with customer lock-in, pp. 84–100 and Appendices G, H, and I.

7. The terminology "deep pocket" hypothesis as coined by Edwards refers to the idea that "an enterprise that is big [has] a special kind of power, based upon the fact that it can spend money in large amounts. If such a concern finds itself matching expenditures or losses, dollar for dollar, with a substantially smaller firm, the lengths of its purse assures it of victory." See C. D. Edwards, "Conglomerates' Big Business as a Source of Power," found in *Business Concentration and Price Policy—A Conference Report of the National Bureau of Economic Research* (Princeton, Princeton University Press, 1955), pp. 334–335.

8. IBM, General Service Agreement Contract, p. A-7.

9. Ibid.

10. Sharpe, *The Economics of Computers,* p. 253.

11. T. G. Belden, M. R. Belden, *The Lengthening Shadow* (Boston, Little-Brown, 1962), p. 309.

12. *Commerce Clearing House Trade Regulation Reports,* 1 (Chicago, Commerce Clearing House), p. 4058.

13. IBM Second Consent Decree, Section XB.

14. Ibid., Section XD.

15. Sharpe, *The Economics of Computers,* p. 255.

16. M. E. Conway, "On the Economics of the Software Market," *Datamation* (October, 1968), p. 30.

17. Sharpe, *The Economics of Computers,* p. 264.

18. *Telex v. IBM,* Northern District of Oklahoma, PX 115, 116, 123, 136, 306, 385.

19. *U.S. v. IBM,* PX 5085.

20. *U.S. v. IBM,* PX 2508A.

21. F. M. Fisher, J. J. McGowan, J. E. Greenwood, *Folded, Spindled and Mutilated: Economic Analysis and U.S. v. IBM* (Cambridge, MA, MIT Press, 1983), pp. 213, 214.

22. See Geruson, "Elements of Strategy in IBM's Price and Product Behaviour," for an evaluation of the development of the plug compatible mainframe processor and plug compatible mainframe peripheral market, pp. 59–100.

23. *U.S. v. IBM,* PX 6747A, 6753A, 6754.

24. This action was originally code named "April" and later became the "Mallard Program"; see Geruson, "Elements of Strategy in IBM's Price and Product Behaviour," pp. 86–88.

25. A disk drive is a medium for disk storage. The basic disk drive apparatus consists of several drive shafts, or spindles, on which disk platters are mounted, read/write heads, and associated electronics circuitry. The controller, as it name implies, controls the flow of data between the processor and the disk drives attached to it. In particular, it interprets instructions from the CPU and directs disk drive activity in response to those instructions.

26. *U.S. v. IBM,* PX 3930.

27. A. Pollack, "Rivals Say Machines Are Not 'Clone Killers'," *The New York Times* (3 April 1987), p. D18.

28. D. E. Sanger, "IBM Offers a Blitz of New PCs," *The New York Times* (3 April 1987), pp. D1, D18.

6

Product Segmentation in the Mainframe Computer Market

The computing market has often been described as a pyramid in which a few extremely large users, including the government laboratories, are at the top of the pyramid. The market broadens down to include aerospace industries and other government supported groups next, then the large commercial users and finally, layers of smaller and smaller commercial users. . . . If one optimizes a computer design on the basis of income, it always turns out that a design aimed for the middle of the pyramid or the lower-middle of the pyramid will result in far greater total income than one tailored either for the top, "Gold Chip" customers, or for the bottom myriad of small potential users where the competition is quite fierce. [IBM's] market philosophy has taken exactly this point of view. [IBM] seems to have optimized their whole compatible structure around total profit, which means they have been optimizing around the middle of the market pyramid.

Dr. Harwood Kolski,
Senior Scientist, IBM[1]

We're not going into peripherals or services or "affordable" super-computers. We're staying with real super-computers and are committed to the top end of that market. The place we've got to watch for competitors is the woods of Oregon, not Tokyo or Armonk.

John Rollwagon, Chairman
and CEO, Cray Research

This chapter shows that the concept of product segmentation is an appropriate heuristic for understanding markets, interpreting behavior, and discerning strategy in primarily the supercomputer niche and also the fault-tolerant computer niche within the mainframe industry. The concept of product segmentation as developed and defined in Part I of this book, draws on the logic of both product differentiation and customer segmentation. But product segmentation is more than simply the exploitation of product heterogeneity and customer heterogeneity in one behavior. It is more than simply the simultaneous use of product augmentation and price segmentation with one activity such as product bundling. Product segmentation, rather, is based on the more complex logic whereby the exploitation of one type of heterogeneity is a vehicle for exploiting the other type of heterogeneity. It is thus segment-based augmentation and product-based segmentation.

The existence and effective use of product segmentation in actual markets requires three basic, related conditions be met: (1) the segmentability of the customer base on the basis of a product need, (2) the augmentation of a product in a way that better meets this segment-specific need, and (3) the ability and willingness to sustain this segment-specific focus despite the lures associated with a strategy directed toward a broader market appeal. The first part of this chapter focuses on describing the empirical basis for both the customer conditions and product conditions required for product segmentation (conditions 1 and 2) in the supercomputer niche.

The concept of product segmentation is not only useful as a descriptive tool for understanding markets and behavior but also as a normative device for understanding the market success and failure of a segment-focused strategy. In particular, in this chapter, the effectiveness of Cray Research in the supercomputing niche and Tandem Computers in the fault-tolerant computing niche will be seen to be related to the ability to meet the requirements of product segmentation in the first place (conditions 1 and 2) and the ability and willingness to sustain a commitment to this segment-focused policy despite the lures of strategy directed at the general market (the third condition). Conversely, the ineffectiveness of segment-focused strategies of other firms will be seen to be related to the failure to meet the requirements of product segmentation in the first place (as is the case with the "BUNCH"[3]), the lack of a clear commitment to a segment-specific strategy from the outset (as is twice the case with IBM), or the failure to sustain the segment-specific focus of a previously effective strategy (as is the case with Control Data Corporation).

Evidence that Cray and Tandem, and their respective market segments, meet the requirements for product segmentation takes the form of a description of customers and products based on publicly available information. Given the conditions for product segmentation, the commitment or lack of commitment to this segment-focused strategy is revealed by both emergent behavior and intended behavior. Evidence for this commitment, or lack thereof, takes the form of observation of actual behavior as well as explicit statements by senior executives, both obtained from court documents, the business press, and case research.

MARKET SEGMENTATION VERSUS
PRODUCT SEGMENTATION

The strategic incentive to focus on a segment of the customer base lies in the potential of such a strategy for isolating a firm from effective general market competitors. This is exactly the case in the mainframe industry. IBM's dominance in the general mainframe market is both a threat and opportunity for rival firms. On the one hand, IBM's highly effective risk-reducing augmentation and customer lock-in-based price segmentation in the general market, and to a lesser extent the more recent low, market-entry pricing of Japanese firms, has forced traditional major IBM competitors to abandon their broad customer base and turn to segment-focused strategies in order to remain viable. On the other hand, IBM's market approach of offering products with the broadest customer appeal has been treated by some new firms as an opportunity to exploit the heterogeneity of the customer base and more precisely satisfy the needs of a small segment in a way that IBM's general market approach precludes.

The result has been the wide use of customer segment-focused strategies.[4] In particular, the prediction of one IBM executive after the introduction of the IBM 360 series mainframe that it is "highly unlikely" that new and traditional IBM rivals "will attempt to compete 'across-the-board' "[5] has proven accurate. Over fifteen years later Sobel concluded that: "The only successes in the information processing field have gone to firms that made oblique assaults, concentrating on special markets and customers, content to live off IBM's leavings. . . . They remain specialized firms, knowing that to lay down a challenge across the board would be disastrous."[6]

To a large extent, the design of segment-focused strategy has been based on the conventional customer-segmentation criteria of marketing. The traditional major IBM competitors, the BUNCH for instance, have defined their target segments on the basis of the industry of the customers. As Magnat summarizes:

The BUNCH companies have been transforming themselves in the last few years from fully integrated manufacturers to marketers . . . directed toward defined niches rather than to the data processing market at large. NCR . . . has focused on the retail and banking markets. Honeywell [has focused on] process control systems for manufacturers . . . Control Data has zeroed in on the scientific and engineering fields . . . Burroughs [has targeted] hospitals, Sperry [has targeted] airlines and government.[7]

Such focused marketing efforts are exemplary strategies of "market segmentation" in the marketing sense. They are not, however, product segmentation as developed in this thesis. As product segmentation, the strategies of the BUNCH are insufficient because they do not fully utilize both dimensions of market heterogeneity. In particular, while BUNCH segmentation strategies do employ customer segmentation, the basis of that segmentation is not the need for a specific product augmentation and relatedly their behavior does not entail significant product augmentation. The focusing of sales efforts on one or two customer-industry sectors may result in some competitive advantage, but it does not necessarily mean customers are being targeted on the basis of a distinct product need or that they are being offered a product that is distinctive in its ability to meet that need.

The inadequacy of the BUNCH strategies therefore lies first in the failure to define customer segments and thus delineate their target markets in terms of product benefits needed. The failure to segment customers on the basis of product augmentation needed, in turn, has resulted in the inability to significantly differentiate their products with respect to this segment. And indeed, the failure of the BUNCH to substantially distinguish their products from IBM is symptomatic of the superficial customer characteristics on which the target segments were based in the first place.

More specifically, despite some specialization in the sales force, the BUNCH remain wedded to the same fundamental architecture and technology that traditionally have well served the general market. As Figure 6.1 illustrates, unlike Cray and Tandem, which have differentiated their products in terms of fundamental design (vector processing and multiprocessing, to be discussed), the BUNCH has retained von Newmann-type scalar processing (discussed later) that best serves the

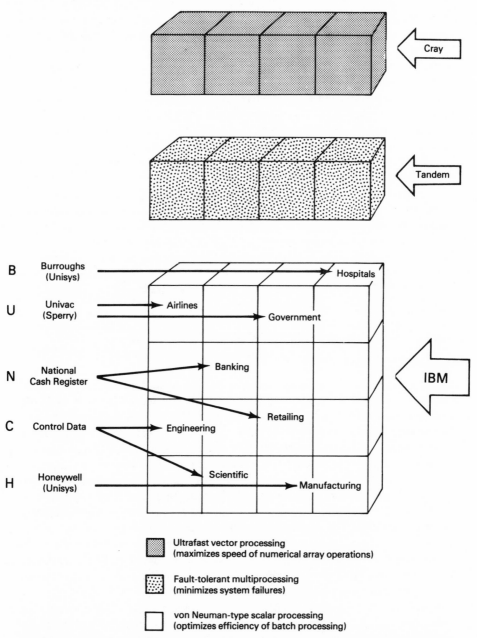

Figure 6.1. Product segmentation versus market segmentation in the mainframe computer industry.

needs of the general purpose market. As will be seen, the distinctive nature of the segment-focused behavior of Cray and Tandem is that they have defined their target segments in terms of a product need and differentiated their product to better match this segment-specific need. It is only when this product-dimensional logic is integrated with the customer-dimensional logic that a segment-focused strategy creates a band of insulation against competitive threats. Moreover, because such product augmentation is done with respect to only a customer segment, rather than the entire customer base, it not only serves as a means of protection against price competition but also against the competitive threat of general market augmentation.

Because of the failure to define target segments in terms of augmentation needed, and the resultant inability to significantly differentiate their products with respect to that segment, the segment-focused strategy of the BUNCH has not insulated them from competitive threats in the way that product segmentation has insulated Cray and Tandem. This, in part, may explain the lackluster performance of the former and the excellent performance of the latter. The next section shows how the basis of Cray's target segment is, in fact, nothing else but a special need for a product augmentation. The section after next shows how Cray has definitively augmented its product with respect to that segment-specific product need.

THE CUSTOMER DIMENSION OF PRODUCT SEGMENTATION IN THE SUPERCOMPUTING NICHE

The first requirement for product segmentation is the segmentability of the customer base on the basis of a product need. The conceptual distinction between customer heterogeneity and customer segmentability discussed in Chapter 3 is important here. Segmentability does not simply require customer heterogeneity but a particular structure to that heterogeneity. Specifically, segmentability requires that along some customer dimension there are significant differences *between* the general market and a segment of the market, while it requires substantial commonality *within* the customer segment as to this customer dimension. Thus the concept of segmentability, applied to product segmentation, requires the presence of a product need that, on the one hand, is significantly distinctive from the bulk of customer needs and yet, on the other hand, is sufficiently similar among a group of customers to constitute a viable target segment. Further, the potential for competitive insulation by augmenting a product to more closely match this need is related to the relative intensity of this distinctive yet common product need.

The following two subsections review the customer condition underlying an exemplary manifestation of product segmentation in the mainframe industry. In particular, the next subsection reviews the customer dimension of product segmentation in the supercomputer niche and, specifically, examines the source of the customer's desire for ultrafast computing as evidence for the presence of an intense need common to a group of customers within the mainframe industry. This evidence takes the form of descriptive customer examples gathered from the business press and case research. The subsection following shows that this product need, although common to a variety of customers, is distinctive in that the number of customers sharing this need is very small relative to the total number of customers in the market.

The Basis of Cray's Target Segment: An Intense Product Need

Because the segment targeted by Cray is based on a product need, its customer base, as will be seen, spans diverse industry types. Hence unlike the BUNCH companies, which have defined their customer segments in terms of industry, industrial sector, or similar variables, customers within Cray's segment cross the boundaries of a variety of industrial classifications, business types, and geographical regions. (See Figure 6.1.) The need, which distinguishes and defines the small group of otherwise diverse customers that has become known as the supercomputer niche, is summarized in a single word that represents one of several possible core performance attributes associated with a computer: speed.

Although the types of customers desiring ultrafast computers are diverse and although the sources of this need, when viewed from the detailed level of specific computer application, are diverse (as will be shown), the sources of the need for speed when viewed at a more general level can be categorized into two groups. To a very large degree supercomputer users' insatiable appetite for ultrafast computing derives from either the desire to simulate physical phenomena or their need to perform time-consuming computation for, specifically, large-scale research and development projects.

A large portion of current supercomputer customers share the desire to mathematically model dynamic physical phenomena and simulate behavior of these phenomena on a computer terminal to better understand, test, and predict real world events. In particular, extreme computing speed is needed to simulate a variety of natural events above, below, and on the earth's surface, at the subatomic level and the astrophysical level as well as other events that do not occur naturally, such as the flow of gaseous or liquid matter around man-made solid objects.

Although the use of the supercomputer for simulation was originally initiated by the U.S. Atomic Energy Commission, which was attempting to simulate nuclear explosions at the time, supply has in a sense created its own demand. The newly found power of the supercomputer in one area awakened engineers and scientists to the possibility of dynamic modeling and simulation in other areas. The customer need was driven not simply by the desire to do the same old tasks faster but the desire to do tasks previously too difficult to do at all or tasks not even conceived. The need for ultrafast computing capability thus not only stems from the customer's desire to perform specific applications that require such speed, but from the seller's demonstration that it can offer this high-speed capability, thereby creating an awareness of the possibility of high-speed computation that drives new customer applications, which in turn intensifies the need for speed.

Among the most immediately practical uses of the supercomputer is the simulation of air flow around solid objects such as cars, space vehicles, and aircraft. The current Chairman and Chief Executive Officer of Cray Research explains the customer benefit received from applying supercomputing to the problem of wing design:

Before, when an aeronautical engineer designed a wing, he or she had to slice it up in cross sections because the computers could only do a two-dimensional analysis of the wing. This meant the engineer had to be a very good mathematician, because dividing a wing into cross sections is not intuitive; it's fairly abstract. Now, with super-computers, it's possible to do a three-dimensional analysis which means you can fly the whole wing in the computer mathematically and obtain three-dimensional graphics. This output conforms to concepts that the engineer is comfortable with and permits a greater play of creative intuition.[8]

Although Cray's supercomputers have made it possible for the first time to simulate air flows around a proposed wing, aeronautical engineers still lack sufficient computing speed to model the flow of air around a complete aircraft to conclude with confidence the design will work. Engineers can crudely approximate the stimulation of a complete aircraft by piecing together separate simulations of wings, tails, and fuselage but, unlike a complete simulation, such an approximation is an adequate substitute for building and testing full-scale prototypes through trial and error.

As a result, despite the enormous benefit already provided by Cray's computers, aircraft design customers still seek more power. This fact assures a degree of secure demand for the proposed Cray-3, which is said to enable the modeling of the behavior of an entire aircraft. What these engineers need is not better service or more computing features but the single product attribute that can solve their problem: speed.

The simulation of natural events is another major source of the need for ultrafast computing. The forecasting of weather conditions, in particular, generates an intense customer need for computing speed. The use of the fastest computers to forecast weather is almost as old as the supercomputer segment itself. Despite the exponential increase in computing speed over the past few decades, however, weather forecasters continue to demand faster and faster computers. The reason lies in the nature of the problem the customers want to solve.

Forecasting involves the division of the earth's atmosphere into a three-dimensional grid, or matrix. Weather stations on the ocean, on land, and in the atmosphere measure wind velocity, temperature, and humidity at each point on the grid and feed the data into the computer. The solution for each grid point influences the forecast for the points adjoining it. This is because weather changes are the chain reaction of individual molecules. Solar energy, the topography of the earth, and the rotation of the earth all affect this molecular motion and are incorporated into the forecasting model.

The accuracy of the forecast is directly tied to computer performance: the denser the matrix, the more accurate the information. Satellites have dramatically improved data gathering and reduced its cost. The bottleneck is calculation speed. The denser the grid, the longer it takes to complete the calculation. Weil suggests that, given the immense set of iterative calculations involved, a high-end, large, general purpose IBM CPU would take longer to calculate the forecast than the time period to which the forecast extends.[9] In other words, if the top-of-the-line IBM computer were used, tomorrow *morning's* forecast would be completed tomorrow *night*. Fishman explains that only recently have meteorologists at the National Aeronautics and Space Administration's Goddard Modeling and Simulation Facility in Washington been able to move from a three-day forecast to a one-week forecast while maintaining reasonable accuracy (not sacrificing grid density).[10] The answer is not more data nor new concepts, but rather, more speed.

The Global Atmospheric Research Program, a worldwide effort to research climatic conditions demonstrated that the usual five-day forecast could be doubled to ten days and still be generated in a few hours. The European Center for Medium-Range Weather Forecasts located in Reading, England, used a Cray-1 to process and assimilate millions of data elements transmitted from observation stations distributed around the globe. The resultant continually updated ten-day forecasts have been exceptionally beneficial to the shipping, agricultural, construction, air service, and energy industries that rely on such forecasts.

More recently, the simulation of natural events under the earth's surface has fueled the need for ultrafast computing. Recent models of Cray's supercomputers have, for the first time, allowed companies to simulate the shape and behavior of oil basins and other seismic activity associated with underground flows of gas, water, and oil. This in turn has helped oil companies increase yields by thousands of millions of dollars. A written report based on an internal presentation assessing Cray's prospects in the oil industry delivered to Cray's top management explains the conditions that gave rise to the application of Cray's supercomputer to oil exploration:

In the early 1980s, oil companies were swimming in cash and searching for attractive investments. Exploration production costs had been increasing rapidly, and reservoirs of new oil discoveries were declining in size compared to previous years. As a result, geologists, geophysicists and reservoir engineers were encouraged to develop more sophisticated analytical techniques to evaluate oil investment projects. These techniques demand significant computational power to obtain reasonable accuracy. As a result, technically advanced reservoir engineers and state-of-the-art geophysicists turned to the super-computer industry to find the needed computational power. . . . Oil companies regarded the Cray as a small exploration expense that paid off handsomely.[11]

The report estimated that "reservoir engineering demand is increasing 20% to 30% per year" and "seismic processing demand is increasing 30% to 50% per year," and concluded "the oil industry's thirst for computational power is large and growing."

The source of the need for ultrafast computing is also manifest in simulations of natural events on a much smaller scale. Ultrafast computing has allowed scientists to unravel the structure of molecules through simulations. Only recently, for instance, has science revealed the structure of the enzyme glycogen phosphorylase, the largest asymmetrical molecule analyzed to date.[12]

But, again, despite the breakthroughs in computing speed, the need of the customers doing this research remains intense, for the longer-term problem they are addressing requires more computing speed. In particular, the longer-range task of these researchers is to understand how enzymes bind to DNA material in cells by modeling their structure and then simulating this behavior—the ultimate goal being the genetic alteration of enzymes to curb the growth of cancer cells—a pursuit requiring more and more computing speed.

The goal of simulating even smaller-scale natural events further illustrates the intensity of this special product need. Simulating the behavior of smaller and smaller subatomic particles in particular may be a bottomless pit for needed computing speed. For instance, using modern quantum chromodynamics, the prediction of the mass of a stationary proton requires roughly three hundred thousand million complex calculations: a task that would take a Cray-1 at least several decades to perform. The result, compared to the actual mass, would provide the first precise test of the theory; as a result, this task has already been assigned to proposed, more powerful, future supercomputers.

Other activities demanding ultrafast computing in the field of physics include magnetic fusion energy research, isotope separation, laser fusion, and research trying to improve the safety of nuclear reactors. Some problems remain beyond the capabilities of current supercomputers thus assuring a continued, identifiable segment of future customer demand. For instance, the completion of the task of creating "a minutely detailed three-dimensional model of a fusion reactor's interior at work" would take state-of-the-art supercomputers hundreds of years.[13]

Other customers of ultrafast computers include automobile manufacturers who use them to simulate crash testing, thereby improving the safety of the cars while reducing the level of expensive, actual testing required. These customers also use ultrafast computing to design engines as well as car bodies, which in both cases, have improved fuel efficiency and, in the latter case, has eliminated much costly wind tunnel testing.

From the very first application of supercomputers to crack codes for the U.S. National Security Agency and to design weapons for the U.S. Navy in the 1950s, the military still remains one of the most hungry users of ultrafast computers. The intensity of this customer need will likely continue especially in light of such projects as the proposed U.S. Strategic Defense Initiative, which presents the enormous computational problem of tracking a myriad of projectiles and calculating optimal interception points for defensive weapons within severe time constraints.

Other customer projects further assure continued intense demand for this distinctive yet shared product need. Customers attempting to understand human speech and develop artificial voice recognition, like those attempting to create "artificial intelligence," believe the solution to their problems lies, in part, in faster supercomputers not yet developed. With the inclusion of producers of computer-generated animated movie scenes to this group of ultrafast computer users, the list is roughly comprehensive.

The Distinctiveness of Supercomputing Customers: The Relative Size of Cray's Target Segment

Despite the fact that the need for ultrafast computing is intense, growing, and common to a variety of customer types, it is distinctive in that the number of customers with this need is very small relative to the total mainframe market. Indeed, the fact that a roughly comprehensive list of Cray's customers and their needs can be described in just a few pages of text highlights the small size of this customer group relative to the total number of mainframe customers, who span the entire range of large organizations comprising modern civilization. The limited size of the supercomputer customer segment is confirmed by Cray's internal market data. In terms of both the total number of customers and the total number of systems installed, Cray's data shows the ultrafast computing segment to be small. By the beginning of 1985, over a dozen years after the company was founded, Cray had only sixty customers worldwide. Although the average number of systems per customer has been increasing, this represented a total installed base of only eighty-eight computers.[14] Yet these eighty-eight customers made Cray the dominant competitor with a 70 percent share of the total market segment.[15]

In 1972, when Cray Research Inc. was founded, the company estimated the total market to consist of 80 to 100 customers worldwide: a customer level Cray has only recently reached. Since then estimates have been revised upward partially as a result of "technological advances that made possible deep price cuts—as much as $5 million to $10 million—and put the machines within reach of more users."[16] Wiel, Morgan Stanley's former chief computer industry analyst, estimates the market potential to be limited to between 80 and 200 customers.[17] Similarly, Stephen McClellan, three times president of the Computer Industry Analysts Group and a vice president at the Solomon Brothers investment bank, estimates that the long-term potential market consists of between 100 to 200 users, of which Cray already serves three-quarters. Moreover, McClellan estimates that about 60 percent of all business in this product segment is concentrated in three customers: the U.S. Energy Department, the U.S. Weather Bureau, and the U.S. Department of Defense.[18] More generous estimates include a *Forbes* magazine estimate of 400 potential customers and an optimistic estimate by Peter Gregory, Cray's director of corporate planning, who sets the long-term potential for the market at closer to 600 users, with about half designated as "probables."

Even the upper range of these estimates, however, represents an order of magnitude that is minor relative to the existing, nonetheless projected, customer base in the general commercial mainframe market. Moreover the small size of Cray in terms of customer numbers does not mislead the reader as to the small dollar size of Cray's segment relative to the general market. The relative smallness of Cray in monetary terms is emphasized by the following comparison: the total *revenue* Cray generated for the *entire year* of 1985 is .05 *percent* of the *net income* generated by IBM for the *first quarter* of that same year.

Hence despite the variety of ultrafast computing applications, Cray's customer base is small both in terms of the relative number of customers and in terms of relative dollar revenue. This small size is important because it minimizes the revenue attractiveness of this segment to general market competitors and to IBM in particular. This in turn allows a small competitor like Cray to establish itself unencumbered by scale-based cost advantages that IBM and other larger competitors might possess.

These last few subsections have highlighted two important features of Cray's segment-focused market posture. The previous subsection showed that, unlike the BUNCH, Cray's basis for defining its customer segment lies not in traditional market segmentation variables, such as customer industry, but in a product need. In particular, it showed that Cray's target group of otherwise diverse customers is defined on the basis of an intense product need that is shared by all members within the group. This subsection showed that, although this need is intense, growing, and common to a variety of customers, it is distinctive in that the number of customers possessing it is small relative to the total market. Both these facts—the intensity of the need and the limited number of customers possessing it—mean that successful product augmentation with respect to this segment can be effective in insulating Cray from unfocused competition in the general market.

Given appropriate customer conditions for product segmentation, the effectiveness of such a strategy hinges on the ongoing ability and willingness to implement product augmentation. The next section reviews the product dimension of Cray's strategy and specifically the emergent product behavior of Cray Research from the time of its founding. Cray's product actions reveal its demonstrated ability to implement segment-based augmentation and further reveal a pattern of ongoing commitment to sustaining its product advantage. Later sections review both emergent behavior and intended behavior of Cray and the major supercomputer manufacturers prior to Cray (IBM and CDC) to show the utility of the concept of product segmentation for diagnosing segment-based strategy and, in particular, the importance of sustaining focus in effective product segmentation.

THE PRODUCT DIMENSION OF PRODUCT SEGMENTATION IN THE SUPERCOMPUTING NICHE

Device Technology

Cray has periodically significantly extended the limits of computing speed through improvements in device technology and changes in computer architecture. Cray has improved device technology through new chip technology, new chip-packaging technology, and new cooling systems for this packaging. Computer chips have been improved by fashioning them with new materials that allow them to operate at faster and faster speeds. Toward the same end, Cray has developed new chip-packaging technology. The object of repackaging chips is to miniaturize circuitry in order to reduce the distance electric signals have to travel, thereby reducing the time they travel and therefore improving speed.

Cray's denser packaging is achieved by shortening the connecting wires between the computer circuits, by rearranging the circuits, or by making the transistors on the computer's integrated circuits smaller so they can be packed closer together on a chip. The greater the density of the circuitry, however, the greater the heat generated by the system, and thus the more powerful the cooling system required. At the same time, the greater the density of the circuitry, the more difficult to engineer a cooling system that does not dilute this density. Cray's augmentation hence also lies in its successively more effective cooling systems.

Early high-speed computers (discussed later) achieved greater speed by replacing vacuum tubes with germanium transistors (the CDC 1604) and then by using silicon transistors that were faster than the germanium type (the CDC 6600). Cray's first product, the supercomputer Cray-1, achieved greater speed through new architecture but also through denser circuit packaging facilitated by structuring the frame of the computer as a series of 9-foot columns arranged to form a 6-foot wide, hollow *C* shape. The dense packaging of Cray-1 circuitry generated so much heat that it had to be cooled by circulating the bubbling liquid refrigerant, freon, through a seeming tangle of tubular channels surrounding the integrated circuit boards.

The Cray-2 improved on the Cray-1's speed by packaging chips in three-dimensional blocks, thereby shortening the average length of the wires connecting the computer's circuitry from 4 feet to 16 inches. The Cray-2 packaged the equivalent circuitry contained in the 6-foot-wide, 9-foot-high Cray-1 structure into an area 38 inches long and 26 inches high. The resultant increase in junction temperatures by 18 degrees Celcius halved the life of integrated circuits and required a more powerful cooling system. Unlike the Cray-1, which was riddled with cooling tubes filled with freon, the entire Cray-2 was immersed in a bath of another inert liquid fluorocarbon that kept the machine at room temperature.

The Cray-3 uses chips fashioned from another semiconductor material—gallium arsenide—that has an electron mobility four to five times that of silicon and is said to improve transistor switching speeds five- to tenfold. Wire connection distances have been shortened even further; the longest connection in the Cray-3 is about 3 inches. It is highly likely that even greater speed will be achieved through the use of superconducting materials virtually eliminating electrical resistance and thereby making further miniaturization possible. Indeed, most recently, superconductivity "scientists at Stanford University are refining two prototypes [of] a micron-thin film that [will] eventually become high speed pathways between computer chips . . . used in the microscopic circuitry of advanced computers."[19]

Computer Architecture

Changes in computer architecture have also been very important in achieving higher speeds. Unlike improvement in device technology, which attempts to extend the physical limits of what a computer can do by changing the *physical* design (materials, packaging, cooling system), improvements in computer architecture attempt to enhance performance by changing the *logical* design so that the computer handles data in ways that result in more rapid solutions to certain types of problems.

The single most important source of Cray's augmentation lies in a computer architecture that allows its computers to perform vector processing. Unlike computers before it, the Cray-1 was not only able to perform calculations one at a time (scalar processing) but also was able to perform many calculations simultaneously through vector processing. Vector processes are distinguished from other computers by their ability to operate on numerical arrays of data, for instance, their capacity for inverting very large matrices or performing highly complex integral calculations (such as Fourier analysis) at extremely high speeds. Since the introduction of the Cray-1, the most important criteria a computer must meet to be classed as a supercomputer is not simply its speed but its ability to do vector processing, which in turn dramatically improves its speed in performing the types of calculations noted above.

The unique design features that permit Cray's computers to do vector processing are multiple parallel circuits called "pipelines" and an instruction set that optimizes its performance (speed) in operations on arrays of numerical data. The key to this speed is the pipeline, which performs calculations on large numbers of vectors (containing uniform and repetitive data) in an assembly-line fashion rather than one at a time. The data flow through the pipeline's specialized processing stations in continuous streams. As long as the pipelines are kept filled their combined output far exceeds that of a scalar-only processor.

The actual speed of a supercomputer depends not only on its vector-processing capabilities but also on its scalar-processing capabilities for computations dealing with single numbers or pairs of numbers that cannot be specified by vectors. This fact makes it very difficult to reduce the performance of supercomputers to a single speed measurement. Commercial general market computers use MIPS (Millions of Instructions Per Second)—a scalar measure—as the common speed-measurement unit. Supercomputers use MFLOPS (Millions of Floating Operations Per Second)— a vector measure—as the common speed-measurement unit.

MFLOPS alone, however, can be an inadequate measure of supercomputer speed. For instance the Cray-1, offering a lower peak megaflop rating than the Cyber 205 (250 MFLOPs versus 400 MFLOPS) but a faster peak MIPS rating (80 MIPS versus 50 MIPS), could outperform the Cyber 205 on most applications because the mix of capabilities was better suited for these applications. Component switching-speeds, machine-cycle times, memory size, compiler efficiency, pipelining capability (having repetitive sequential steps performed simultaneously), and peripheral-device capabilities also affect a supercomputer's speed. The result of this complexity is that supercomputer speeds can be reliably compared only for specific applications performed in trial runs.

Another source of increased speed lies in the more radical design approach called parallel processing or multiprocessing, which involves linking any number of independent processing units, or nodes, in parallel to speed the execution of a single program. Until very recently, all commercially successful mainframes, the Cray-1 included, were based on von Newmann architecture in which the computer processor is entirely distinct from its memory. This separate memory contains not only the data for a particular problem but also instructions for manipulating that data. The central processor receives pieces of data and instructions from the main memory step by step, pausing after each step to send the results back to the memory. All the information flowing back and forth between the processor and memory passes through the same, single electronic pathway that has become known as the "von Newmann bottleneck."

Instead of advancing in a sequential step-by-step manner, parallel processors achieve greater speed by dividing up a problem and processing its parts simultaneously on multiple computing modes. In other words, instead of funneling information through what amounts to an "electronic straw" (the von Newmann bottleneck), parallel processors soak up data like sponges. Experimental parallel processors have ranged from the "coarse-grain" type in which only a small number of very powerful processors are linked to "fine-grain" types in which a very large number of weak processors are linked (a design that crudely approximates that of the human brain).[20]

Cray has proven the parallel design approach commercially viable with the market success of the Cray X-MP, the first supercomputer incorporating elements of parallel processing design. Unlike the Cray-2, which achieved greater speed through denser packaging and the miniaturization of integrated circuit boards, the Cray X-

MP achieve greater speed by linking together Cray-1 type processors: two processors in the case of the Cray X-MP-2 and four processors in the case of the Cray X-MP-4. These processors are synchronized through clusters of shared registers in the CPU intercommunication section and through central memory.

It is not clear currently whether device technology or computer architecture is a more potent source of speed increases. (Cray is pursuing both avenues.) Eventually, however, greater speed will come more and more from improvements in computer architecture. Despite dramatic speed improvements from new materials, packaging techniques, and cooling systems, the potential for speed improvements in uni-processors with the von Newmann bottleneck ultimately is more limited than the potential of multiprocessors because of a fundamental physical barrier: the speed of light.

The net impact of Cray's efforts in both computer architecture and device technology is a sustained, ongoing augmentation of its product along the one product attribute where its segment-specific appeal lies: speed. Unlike less tangible forms of augmentation. Cray's ongoing segment-specific augmentation is readily confirmed through the standard quantitative measure of supercomputing speed. As Figure 6.2 shows, since the introduction of Cray's first computer, with every new product introduction Cray has redefined the limits of ultrafast computing.

The concept of product segmentation is not only useful for understanding the measurable market success of Cray (quantified at the end of the next section) but also in understanding the failure of segment-focused behavior by other supercomputer manufacturers prior to Cray. The next few sections of this chapter use the concept of product segmentation to understand the effectiveness or ineffectiveness of segment-focused behavior, and thus diagnose strategy, in the market niche for supercomputers.

THE HISTORICAL EMERGENCE OF SUPERCOMPUTING: MARKET SUCCESSES, MARKET FAILURES, AND THE IMPORTANCE OF COMMITMENT TO FOCUSED BEHAVIOR

The origin of the supercomputer market segment is a record of failures to mix a general market strategy with a focused product-segmentation strategy. The early history of the supercomputer market niche is marked by two failed attempts of the dominant general market competitor, IBM, to sustain an effective focused strategy in the ultrafast-computing customer segment. The first attempt was a genuine effort directed at becoming an effective competitor in this niche. The intention of the second attempt, however, was not to establish a profitable business in this customer segment, but merely to establish a presence in this high-technology segment, irrespective of profitability, so as to sustain and reinforce success in the far more important general purpose market.

The early history of the supercomputer market niche is also marked by the failed attempt of an initially successful focused-product segmentor, CDC, to move into the general market dominated by IBM. This resulted in not only a failure to penetrate the general market but a loss of competitive advantage in the supercomputer niche.

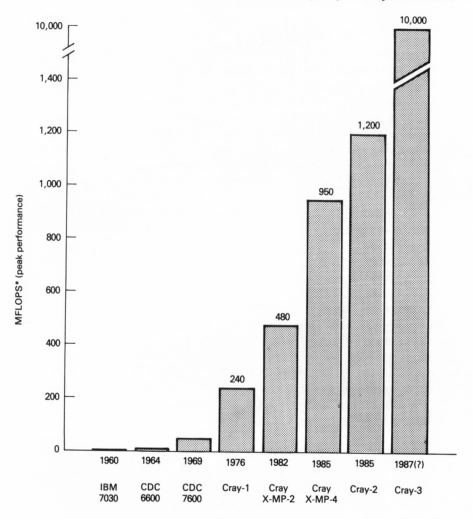

*Millions of Floating (decimal point) Operations Per Second.

Figure 6.2. Improvements in Maximum computing speed available. *Source: High Technology* (May, 1984); *High Technology* (February, 1987); Cray Research.

CDC was effectively squeezed between the high-end general purpose computers of IBM and the faster supercomputers of Cray Research.

Cray's entry into the supercomputer niche has been followed by its sustained dominance of the segment due to a refusal to move into the general market and a constant focus of efforts toward maintaining its competitive advantage among the small number of customers it targets. This section reviews the failed attempts of a general market competitor (IBM) to sustain a segment-focused strategy, the failure of an effective focused competitor to compete in the general market (CDC), and the measurable success of a focused competitor in that same segment (Cray).

IBM's First Failure: The IBM 7030

The origin of the "supercomputer" market segment was customer driven. Seeking ultrahigh speed computing power for nuclear research, the U.S. Atomic Energy Commission (AEC) naturally turned to the largest and most reputable computer manufacturers at the time, including IBM and Sperry Rand. In 1954, the AEC's Lawrence Livermore Laboratory requested bids to build an advanced, ultra-high-speed computer. The following year the AEC's Los Alamos Laboratory contracted with IBM to share in the design of an ultrafast computer, which became known as "STRETCH" (later renamed the IBM 7030). The term STRETCH reflected IBM's expressed intention of stretching the state of the art in technology to produce the world's fastest computer.

IBM set about "exploring the unknown and rethinking and redesigning almost every aspect of earlier IBM computer systems."[21] IBM's computer was to be 100 times faster than its then most powerful commercially available system (the 704). The key to achieving such dramatic augmentation for this very narrow customer segment lay in what was then a new technology: the transistor, which "was faster, smaller, used less power, was more reliable and . . . inherently less costly" than the vacuum tubes it was replacing.[22]

Specifically, the basis for the augmentation lay in second-generation transistors and a unique architecture. The design contained a "look ahead" feature, which allowed the computer to examine and set switches for the next set of instructions while performing operations on a present instruction set. Combined with the new technology, this dramatically improved speed.

However, IBM management at the time, as well as business historians later, dubbed STRETCH "a disaster and a great embarrassment."[23] Inadequate commitment and focus of technical, managerial, and financial resources to the project caused the final product to fall substantially below originally specified performance targets. IBM's failure to meet promised performance capabilities was possibly also partially attributable to the possibility that IBM rushed its development effort to meet competition from Sperry Rand, who was building a supercomputer called the LARC (Los Alamos Research Computer), which was named after its intended customer.

Falling far short of expectations, the first 7030 was delivered to Los Alamos in 1961 at the contracted price of $13.5 million. Due to its unexpectedly low performance specifications, IBM was forced to cut its price to $8 million in order to sell it to the next fifteen potential customers it had lined up.

Given the low price necessary to attract customers, the small size of the potential customer base, the high cost of R&D and production, and thus the low returns relative to foregone investment opportunities in the general commercial business market, IBM concluded that, for a firm in its market position, catering to this segment was a suboptimal strategy, IBM took the computer off the market. It refused further orders for the special machine and agreed to build the machine only for orders previously committed so as not to damage its reputation. In the end, only seven 7030s were completed and installed. The net loss on the project was over twenty million dollars.

CDC's Early Success: The CDC 6600

Two years after the AEC first requested bids for an ultra-high-speed computer, Control Data Corporation was formed with the expressed intention of filling the needs of this customer segment. CDC anticipated its product "would be unique to a great extent . . . it being so much more powerful and so well-suited to scientific work, it would just be outstanding in the eyes of those laboratories that have these very large scientific problems."[24] In their sixth year CDC introduced the 6600, three times faster than IBM's now-dead STRETCH, and the fastest computer in the world at the time.

The basis for augmenting the computer power to such a degree rested in new technology, the genius of the chief design engineer (Seymour Cray), and very importantly, the willingness to focus efforts on types of performance that only a small segment of customers needed at the price of foregoing or even neglecting performance needs of other customers. CDC, unlike IBM, made no attempt to make its machine useful to all users. CDC Vice President Gordon Brown describes a very focused market strategy that optimized performance for a select group of scientific users at the price or reducing its general utility to the vast bulk of customers. He explains:

The strength of the 6000 product line prevailed . . . over IBM and Univac in most [scientific] environments; and, on the other hand, proved to be deficient when it was employed in an environment requiring a lot of input/output of data, or commercial type requirements [because] the architecture of the 6000 series was designed with the scientific user in mind.[25]

Speed was not the only aspect of CDC's strategy. As Chapter 2 suggested, augmentation potential should be viewed from both sides of the customer perceived net benefit equation. CDC, in fact, also reduced the product's cost by eliminating aspects of the augmented product that were exceptionally low in value to this particular customer segment. Specifically, CDC trimmed the customer support component of the augmented product.

CDC's prospectus stated that it did not intend to compete head-on with IBM and other industry giants at the time. Rather it would focus on the special needs of the Department of Defense, aircraft manufacturers, and universities who had unique power requirements, who did not need heavy support services, and who did not require the efforts of a sophisticated sales force to make a purchase decision. This customer segment knew quality when they saw it and knew how to use it and maintain it. CDC's recognition of the unusually high level of sophistication in evaluating and using computer systems among supercomputer customers and, relatedly, their lower perceived purchase risk gave CDC a competitive lever to use against IBM.

IBM's sales force and marketing policy was completely orientated toward the general purpose market of less sophisticated, more risk-conscious, business users. As shown in Chapters 4 and 5, IBM appealed to this general market by augmenting its actual hardware with various forms of support, service, and installation assistance. In this way IBM offered a complete system and a total solution to a customer's data processing problems. This is precisely what the general purpose business user desired. He did not want sophisticated intraproduct comparisons or technical knowledge; he merely wanted a reliable system with minimum risk to perform basic required tasks. This highly effective total-solution selling approach was, of course, also costly to IBM. In the general market, the benefits of this approach, in terms of satisfied customers, more than justified its costs.

IBM used this same approach in its marketing policy for supercomputers. Using its existing sales force, which was geared to marketing "solutions" to unsophisticated buyers, ostensibly made good economic sense. The salespeople were already highly trained and had proven to be effective in the general purpose market. For highly sophisticated supercomputer users, however, such a total-solution marketing approach provided minimal return. As one IBM executive explained, such customers "are much more apt to be influenced by a straightforward scientific discussion of the technical"[26] aspects of the product. Technically proficient buyers found IBM's sales representatives' efforts unnecessary at best and sometimes counterproductive.

CDC's approach, on the other hand, was to allow scientists who actually had a part in developing the computer sell it to the customer. CDC did not have a trained sales force per se; its sales force was its scientists. CDC did not sell solutions; it sold only hardware. It simply explained how to operate the computer and withdrew from the scene. CDC provided "a relatively small amount of software with the customer doing his own software."[27] Support, education (initially), and systems application design were also all left up to the customer. Among the small number of supercomputer customers, this was all that was wanted. Supplemental software and education made little sense because the seller had less understanding of the use to which the supercomputers were put than did the buyer.

Not only did this strictly hardware-orientated, informal selling approach better satisfy potential customers, but it also provided a cost advantages. It was substantially less expensive than IBM's policy of training a group of nontechnical marketing representatives to sell supercomputers. CDC's cost advantage was further enlarged by IBM's policy of providing on-site customer-support services. IBM's policy was to provide a cadre of highly specialized, and thus costly, applied scientists to support installation activities especially at the more prestigious account sites—a practice that was not only redundant but possibly contrary to the wants of customer scientists.

CDC's policy of using technicians and scientists with no sales training to sell its computers was effective given a customer group for which basic support is redundant and speed is of overriding importance. Such a sales tactic is the logical result of a strategic approach that is focused on exploiting the distinctive nature of its customer segment to the fullest in order to form the most precise match between the customer and product offered. William Norris's (CDC's founder) testimony in the U.S. Justice Department case against IBM is accurate: CDC was very successful initially because "we picked out a particular niche in the market"—the "scientific and engineering part of the market"—and "met the needs of the particular part very proficiently and much more so than any computer then available."[28] As a result, CDC's product "enthrall[ed] that small, elite group of customers like the Atomic Energy Commission and the United States Weather Bureau whose thirst for computing power was nearly insatiable,"[29] and the company made rapid inroads into the aerospace, education, and large government laboratory markets.

IBM's Second Failure: The IBM 360 Model 90

Despite its previous failure with STRETCH, IBM reacted to CDC's success by making a second attempt to compete in the supercomputer niche. There are three not mutually incompatible explanations for IBM's second effort. First is simply the somewhat irrational motive of repairing IBM's damaged pride. Second, as defense witnesses in the *U.S. v. IBM* case argued, IBM felt compelled for patriotic reasons to make a more substantial contribution to what was, at the time, a quite explicit national U.S. policy of technological achievement. The most compelling explanation, however, is that of the cold logic of market strategy. Although it had become clear to IBM management that investment in this customer segment could not be justified in terms of its profitability relative to the profitability of the general market, it could be justified in terms of the impact of dominant presence in the highest technology customer segment had on the larger middle- and low-technology portions of IBM's market.

That damaged pride played a role in IBM's reentry is supported by the apparent shock IBM's management expressed (after CDC announced the performance specifications of the 6600) at how a small start-up company, with far fewer human and financial resources, could achieve what IBM had been unable to achieve. Tom Watson, Jr., (IBM Chairman at the time) made clear his sentiments in a notorious "janitor memo," which, in essence, suggested that the ability of a small company to produce a more powerful computer than IBM was a direct slur on IBM's virility. Watson published a revised version of the letter:

Last week CDC had a press conference during which they officially announced their 6600 system. I understand that in the laboratory developing this system there are only 34 people, including the janitor. . . . Contrasting this modest effort with our own vast development activities, I fail to understand why we have lost our industry leadership position by letting someone else offer the world's most powerful computer.[30]

The political climate of the times also contributed to the decision. Because technological advance was a national goal and IBM was the largest computer technology firm, efforts in building a supercomputer were a matter of patriotism. IBM's director of scientific computing explains during testimony:

At that time in history, the President of the United States and the people at large had dedicated themselves towards a substantially higher level of scientific and engineering and technological achievement than the country had experienced prior to that time due to a variety of considerations, including the Russian success in areas of technology and science, and a national goal had been stated relative to the need for the country to achieve great leaps forward in various areas of science and technology.[31]

Watson and others expressed concern that IBM was not responding adequately to the needs of the U.S. government for advanced computing systems in connection with the government's high priority programs in "atomic energy research, weapons development, space exploration," and related areas.[32]

Neither pride nor patriotism nor even "technological fallout,"[33] however, was likely sufficient to induce IBM to enter this market. Even if IBM could compete profitably in this segment, entry could not be justified on a simple profit basis because the return on effort directed at the supercomputer segment was substantially less than returns from similar levels of effort in the general market.

However, entry could be justified from a strategic viewpoint. In particular, IBM reasoned that the impact on customer perception in the general market of a dominant IBM position in the "high-technology" segment would more than justify such an investment. IBM executives correctly recognized that IBM's success in the general market was, in a large part, attributable to the IBM name—the industry leader— because customers were not simply buying a computer but "IBM" itself. IBM's Dr. Harwood Kolsky warned fellow IBM executives that "IBM, by not having a STRETCH like pace-setter in the market, has lost a great deal of this psychological advantage in the computing industry." He argued that this competitive advantage in customer perception could be "largely . . . taken over by the CDC 6600 . . . which gives [CDC] the prestige of being the pace-setter in the computer industry."[34]

IBM consequently decided to enter the supercomputer segment not because of its profitability but to protect the profitability of the general commercial market. Indeed profits in the supercomputer niche were not the goal of the strategy; rather, as Kolsky explained to IBM executives, the IBM entry "should deliberately be done as a competition stopper . . . it should be deliberately be done as a money-loser (or more tactfully, a shared-cost development for the benefit of Government)."[35]

IBM was particularly concerned about the inroads CDC had made into one of the most prestigious gold-chip customers, the Atomic Energy Commission. The AEC and similar customers were important because although they "may not represent a particularly *profitable* market . . . they represent an extremely *influential* market. As they go, so go a great many of the less sophisticated but more profitable customers."[36] Because the AEC was a highly visible customer—a pacesetter— whose choice of computing equipment greatly influenced the decisions of many other important customers, IBM was alarmed that the commission deemed CDC 6600s "more modern and [more appropriate for] the standard scientific problems of the AEC than . . . the IBM product line which they feel is now becoming expensive and obsolete technologically;" and although the AEC laboratories previously had "always . . . shown a preference for IBM equipment," explained Kolsky in a memo to Watson, "almost all of the AEC laboratories presently have competitive equipment installed or on order, much of it earmarked to replace IBM equipment."[37]

Unable to compete in this segment on a profitable basis, IBM nevertheless entered with the intent, not to be profitable, but to sacrifice losses to eliminate the CDC threat. The unusually explicit nature of this intended anticompetitive behavior was reflected by management's reluctance to give their signature of approval to the extremely low price of the 360/90 (IBM's new supercomputer). Several management levels within IBM refused to sign off. Faw argued that "in light of its possible effect on CDC" and its low price, it could never be presented as a fair and competitive action. IBM's Product Testing Department attempted to discourage Watson from going ahead with the product introduction because they had not yet clearly determined its actual "performance or functional specifications." This put IBM salesmen in the potential position of unintentionally or intentionally misrepresenting its performance. Finally, IBM's Chief Counsel, H. W. Trimble, refused to approve of the introduction because "the announcement of the Model 90 is motivated primarily for competitive reasons."[38]

Freed from profit constraints, the price of the 360/90 relative to promised performance made it an instant market success. Unable to match announced IBM price/performance, CDC could not book a single order for eighteen months after the IBM announcement. Attempting to explain this sudden market deterioration, CDC's 1965 annual report complained that "our major competitor . . . is making a highly concerted effort to hinder our progress by frequent announcement of changing characteristics and new models at reduced prices of large computers reported to be under development."[39] Shortly after the 360/90 introduction, IBM's Management Review Committee concluded that competition had been effectively limited to "trading share among themselves."[40] The cost of this tactic, however, was enormous. IBM had lost $126 million for the design and the building of the 360/90s which, in fact, fell substantially short of performance specifications. IBM lost over $100 million more in assets when it was forced to give CDC its entire Service Bureau Corporation in return for CDC's withdrawing the antitrust suit it filed shortly after the 360/90 introduction.

Although IBM's pricing strategy was initially a large success, once supercomputer customers discovered that actual performance was substantially less than announced performance. IBM's market success evaporated. So despite the fact that the 360/90 was freed from the normal economic success criteria (profitability), it still failed to achieve the easier goal of "market presence" because of its inferior performance. Supercomputer customers need speed more than low price. Regardless of how low the product was priced, if it could not do the job, it would not be purchased for the job. In the end IBM built only seventeen 360/90s and sold only thirteen, while CDC went on to sell close to one hundred 6600s. After this second failure, IBM terminated its plans for a follow-on supercomputer project—the Advanced Computer System program headed by Gene Amdhal—and withdrew from the market, never to return.

How is it that IBM, with the hindsight of the STRETCH failure and the resources of one of the most profitable companies on earth, failed to produce a product of adequate performance? The answer lies in IBM's strategic constraints relative to the limits of technology. CDC, by focusing on the supercomputer niche, was free to use a computer architecture that was the most effective in maximizing speed. IBM failed to recognize the need to use customer segment-specific computer architecture and/or decided not to use such computer architecture because of its incompatibility with IBM's other general market systems' software. As John Opel (former IBM Chairman) explained at the time of the product's introduction: "IBM is developing an ultra-high performance system Model 90. . . . The objective is to produce a system which achieves new levels of performance and *is capable of operating efficiently on programs written for other models of System/360.*"[41] (My emphasis.)

The requirement of producing a computer that was compatible with IBM's entire line and capable of running all existing general application software severely constrained the ability to stretch technology to reach new performance-speed specifications. In this way, the attempt to make this IBM supercomputer fit into the existing IBM general purpose product line diminished its chances for market success. IBM's attempt to mix a focused segment strategy with a general market approach was a strategic error that led to the failure of IBM's actions not only in financial terms (as anticipated) but in strategic terms (establishing a presence in this high technology part of the market).

CDC's Loss of Focus: Losing Sight of the Reasons for Success

Just as IBM had tried to extend its general market product into the supercomputer niche (the 360 series), CDC attempted to move from a highly focused segment strategy into the general market. CDC Vice President Brown summarized CDC's new product strategy in explaining that the "product line . . . was evolving to a . . . better balanced product line between both the scientific and commercial users."[42] CDC Chairman Norris testified that the company was attempting to expand its share of business among supercomputing customers to include revenue generated by their general-purpose computing activities:

We found that there were large companies who, while the majority of the work that they wished to do was of an engineering and scientific nature, still they had a certain amount of business data processing. So, we set about to broaden . . . the 6600 so that we could meet the requirements of those customers where the bulk of the work was still scientific but still the 6600 would do the business data processing well enough so that the customer only had to have the one computer . . . as opposed to having two computers, one for scientific and the other for business.[43]

CDC's follow-on product to the 6600, the CDC 6700, was an updated supercomputer whose potential maximum speed was diluted by the requirement that it also be useful for general-purpose commercial applications. The result was that CDC increased its vulnerability in the supercomputer niche to competitors willing to focus exclusively on high speed (at the expense of making machines that were useless to general market commercial users) without creating a CDC competitive advantage in the general market. With the introduction of the Cray-1 Serial One in the mid-70s by Cray Research, Inc.—a supercomputer that was useless in the commercial market but that was significantly faster than the CDC 6700—CDC found itself, inevitably, squeezed between the superior commercial utility of IBM mainframes in the general market and the superior speed of Cray's fully dedicated supercomputers in the supercomputer niche. CDC's market dominance quickly evaporated. According to CDC salesmen few, if any, 6700s were sold after Cray introduced its first computer.

The failure of CDC's strategy lies in the fact that it lost sight of the source and conditions of its initial success. The closer CDC moved to the general market, the more it removed the insulation from general market competition, specifically the protection form IBM, that a segment-focused strategy offered. Moreover, to the extent that CDC's market-broadening efforts encroached on IBM's territory, CDC incited IBM to meet the threat of CDC's diluted supercomputers with speed-enhanced general purpose IBM mainframes. Furthermore, CDC's compromise of

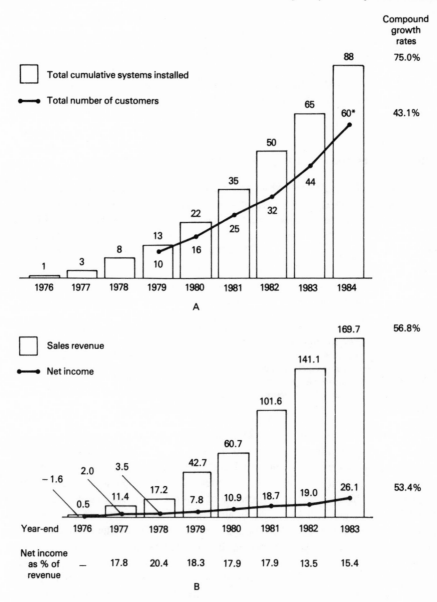

*Extrapolated from 1983 ratio of customers to installations.

Figure 6.3. Market growth and profitability of Cray Research. *Source:* Cray Research.

speed for the sake of greater general market utility made the customer niche for the fastest computers all the more attractive to potential entrants.

Recognizing the technological compromise CDC's top offering had made (and thus the fact that faster computers were immediately technologically feasible) and frustrated at CDC's refusal to fund a proposal to make a faster, more dedicated supercomputer, the chief design engineer at CDC, Seymour Cray, left the company to start his own firm. In 1972 Cray Research, Inc., was formed with the singular objective "to design and build . . . a more powerful computer than anyone now has."[44]

In less than five years Cray had accomplished his objective. In 1976 the Cray-1 Serial One was introduced and installed in Los Alamos National Laboratory in New Mexico on a trial basis. It was found to improve the speed of the CDC 6700 fivefold. The consequent market success of Cray is clearly measurable in terms of volume growth and profitability.

Following Cray's first sale on 1 January 1976, the company entered a period of rapid growth. As A of Figure 6.3 shows, total installations increased at a compound annual growth rate of 75%; calculated from a later starting point, total customers increased at over 40% annually. With over 70% of the market Cray became and today remains the dominant competitor in the ultrafast computer niche; CDC with 22% of the market is the only other competitor of significance, leaving only 8% of the remaining market to be shared among all other supercomputer manufacturers. As B of Figure 6.3 shows, this market share success has also been accompanied by excellent financial performance. For the first six-year period from the end of 1977 to the end of 1984, both profits and revenue increased at a compound annual growth rate of over 50%. For that same period Cray consistently posted a return on total revenue between 13% and 21% annually.

THE DANGERS OF DIMINISHED FOCUS

The market activity in the supercomputing niche prior to Cray's entry reveal the dangers of an inadequate or unsustained commitment to the segment-focused strategy. The reason IBM could fail, where a small start-up like CDC succeeded, lies first in IBM's unwillingness to divert adequate resources to the project (because they would be used more productively in the commercial market) and, relatedly, the unwillingness to forego the compatibility requirement that its product be "capable of operating efficiently on programs written for other models of System/360." Similarly, CDC demonstrated an inadequate commitment to sustaining its segment focus. A key factor in CDC's later loss of product advantage in the supercomputing segment was its actions to "broaden" and "better balance" its product so that it "would do *business* data processing well enough so that the customers only had to have the one computer."

Cray's sustained dominance in the supercomputer segment, on the other hand, is rooted in an effective product-segmentation strategy characterized by an ongoing, unswerving commitment to its customer segment. The unequalled speed of Cray's products allows it to insulate itself from competition, engender customer loyalty, and virtually eliminate product substitutability in the same way IBM's less tangible forms of product augmentation (i.e., support, reliability, and stability) provide IBM with differential advantage and competitive insulation in the general market. Moreover, the inherently small size of Cray's market segment servest to reduce the risk of competition from general market firms whose strategic efforts may reap greater returns when directed at the vastly larger general market.

A danger to Cray's effectiveness, and the mistake that CDC made, is to reduce the competitive insulation created by segment-specific augmentation in an attempt to go beyond the limited sales potential of the target segment. Product actions or sales efforts such as extending or broadening the product line, or simply altering the central product offerings so as to leave a wider market appeal, can strip away this competitive insulation in several ways. First, such product actions can be interpreted by IBM as a direct threat to its share of the general market (whether or not this is, in fact, the case) thereby provoking retaliatory action. Second, to the extent that such product actions are effective in broadening Cray's market, they can increase the likelihood of entry by IBM or other competitors by making the market appear more attractive in terms of revenue potential. Third, changes in the primary-product offerings that sacrifice speed to greater general market utility can directly dilute the basis for Cray's market effectiveness and competitive advantage in the first place. Fourth, even without changing the primary product offerings, the extension of the product line to reach more potential customers in the lower end of Cray's segment and the higher end of IBM's market can distract efforts and diminish resources available for sustaining and enhancing Cray's central differential product advantage. Finally, market-broadening behavior can move Cray into business organizations that have extensive and well-established relationships with IBM (or other general market competitors). The installation of supercomputers to these blue-chip IBM customers can create a strong incentive for IBM, the principle corporate supplier of these customers, to extend its share of its customers' computer spending to include the customers' spending on top-of-the-line supercomputer applications.

To summarize, customer-focused or product-focused behavior that broadens the segment can promote competition from IBM in the general market either directly by provoking a reaction to a perceived market share threat or indirectly by enhancing the attractiveness of the segment to IBM. Further, such product behavior can reduce Cray's ability to compete for its traditional customers, thereby increasing its vulnerability to new and existing supercomputer manufacturers either directly as a result of changes in its traditional products or indirectly through the reallocation of effort and resources away from sustaining its high-speed superiority. The danger of market-broadening behavior by a product segmentor is hence its tendency to invite competition at the low end, high end, or both ends as was the case with the behavior of CDC.

CRAY'S COMMITMENT TO SEGMENT-FOCUSED BEHAVIOR

The review of Cray's product actions in "The Product Dimension of Product Segmentation in the Supercomputing Niche" revealed a pattern of consistent and periodic efforts to push the state of the art in ultrafast computing (recall Figure 6.2) and thus provided initial evidence for Cray's commitment to segment-focused product augmentation. What that section did not reveal, however, is Cray's willingness to pursue greater speed at the expense of computer capabilities in other areas and, more specifically, in areas that could broaden the product's appeal and thus increase *potential* revenue. The lure of such a product/customer-broadening strategy is especially strong for a supercomputer manufacturer because its competitive advantage rests solely on one very specific augmentation: speed. Widening a product's appeal not only increases the potential revenue base but also reduces the high level of dependency on superiority in speed and therefore broadens the basis for competitive advantage. The rest of this section provides further evidence for Cray's commitment to a segment focus and, specifically, its commitment despite the enticement of reduced risk and greater revenue associated with a market-broadening strategy.

Cray's commitment is evidenced by its willingness to sacrifice the product attributes of reliability, compatibility, and general purpose utility to speed. Cray's priority on speed over reliability is explicitly articulated by Cray's Chairman and CEO, John Rollwagon—"if a machine doesn't fail every month or so, it probably isn't properly optimized"[45]—a compromise Rollwagon says his customers willingly accept. Compatibility with general purpose software, and specifically IBM operating systems, is similarly forsaken for speed. This sacrifice is necessary because, as Alexander summarizes the consensus among computer scientists, "operating systems, good as they may be for general purpose business applications, cripple peak performance in scientific super-computers."[46] These sacrifices, as well as others (such as a limited ability to attach to various peripheral devices), constrain the range of uses of Cray products (making them less desirable to the vast bulk of mainframe customers) in exchange for the ability to provide extraordinary, highly specific utility to a very small segment of customers.

The lower general utility of Cray's product is confirmed by the lower per unit revenue Cray's computers earn compared with that of IBM's computers. A comparison between general purpose and segment-specific high-speed computers at a given point in time is helpful in illustrating this. Although the speed measurement units are different for the two types of mainframes (MIPS versus MFLOPS) and therefore not precisely comparable, they are useful for a roughly accurate picture of relative price/performance levels.

Figure 6.4 summarizes price and computing speed information in terms of price/performance slopes. (The steeper the slope the higher the relative price.) From this exhibit it is immediately apparent that the IBM products, which are priced on a constant price-performance level, have a much higher relative price than Cray's products. In particular, the Cray-1 provided not quite 27 FLOPS per dollar (240 MFLOPS/$9 million), while the most powerful IBM computer at the time, the IBM-3081, provided only about 3 MIPS per dollar. Adjustment for differences in the speed-measurement criterion would only widen the disparity in these ratios. IBM's computer is significantly more expensive relative to its speed.

Similarly, a later Cray computer, a version of the Cray X-MP, provided over 52 FLOPS per dollar (630 MFLOPS/$12 million) while the most powerful IBM computer at the time, the IBM-3084, provided again only about 3 MIPS per dollar. (Note the rationale for the constant IBM price/performance slope is IBM's desire to prevent product cannibilization among compatible computers within IBM's product line.) IBM's latest most powerful computers, the IBM 3090s (the "Sierra" series), not illustrated, which in fact are the first IBM computers to incorporate elements of

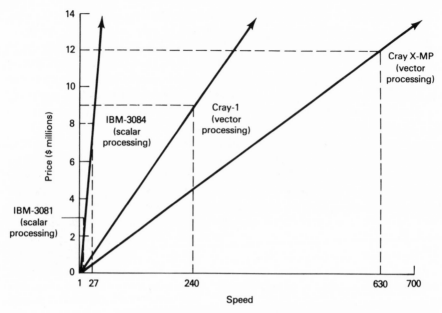

(Millions of instructions or floating operations per second)

Figure 6.4. Prices and speed of general and segment-specific mainframe computers. *Source:* IBM, Cray Research.

vector processing, also give substantially less speed per dollar at their lower $6 million price than do the latest Cray computers priced at $18 million.

This higher price per speed unit reflects the greater general utility of IBM's products. Because the IBM machines operate in a scalar mode, they are extremely versatile in addressing a broad variety of commercial uses. Moreover, because of their high power relative to other non-Cray products, they also can be used in a wide range of less demanding scientific applications and thus are ideal as a multipurpose business computer.

Cray's computers, on the other hand, are vector processors dedicated to optimizing speed in the very specific types of operations. If these special purpose machines were used for more typical, nonvector processing (general purpose business applications such as payroll), they would provide no meaningful advantage. Moreover, Cray's computers often have no utility at all for such more usual applications due to incompatibility with common business software that derives from Cray's unique architecture. Cray has therefore forsaken providing utility to the vast bulk of the customer base in order to provide the greatest possible utility to a small segment of that base.

Cray's commitment is demonstrated not only by its unwillingness to compromise its segment-specific product augmentation in its primary product offering but also by the steadfast unwillingness to broaden the product line in order to appeal to customers beyond its limited niche. A major instance of this ongoing periodic rejection of product behavior that would tend to generalize its market occurred in the early 1980s when the company had under serious consideration the development of a computer less expensive and slower than it every previously offered. The objective behind the proposed computer, dubbed "Quarter Horse," was to expand Cray's market in the low end where the distinction between IBM's customers and Cray's customers became less clear. But Cray Research concluded that it should let new start-ups in this market "fight it out among each other and with IBM." It was decided that Cray was "not going into . . . affordable super-computers" but would stay "committed to the top end of that [supercomputer] market" and product only "real super-computers," which demand "being technically the best."[47]

Urlich Weil, former chief computer industry analyst at the investment bank Morgan Stanley, suggests Cray's ability to avoid the trap of losing focus derives, in part, from the structure, culture, and management style of the Cray Research company.

Mr. Cray's decision to pursue the large-scale scientific market exclusively, despite what some perceived as rather limited market potential, has kept management's attention riveted on meeting the challenges of this demanding set of customers. By keeping the company small, cohesive, and devoted to one specific task, management has avoided the design and manufacturing delays often caused by stifling bureaucracy and overlapping authorities. At the same time, the dedicated efforts of a small but highly qualified design team resulted in very good productivity at a surprisingly reasonable cost . . . Cray had many times been urged to broaden its product range to include medium-sized scientific computers as a means to enlarge the market served. Management has resisted this temptation, preferring controlled, highly profitable growth to a diversion from what the company does so well . . . the company has little to worry about, in our view, provided that it adheres to the principles on which it was founded.[48]

Beyond these qualitative intraorganizational variables there is observable quantitative organizational evidence that Cray has concentrated its efforts in sustaining and improving the effectiveness of its segment-specific augmentation while foregoing potentially greater revenue associated with market-broadening efforts. The distribution of its marketing resources and R&D resources suggests that Cray is concentrating on customers only interested in speed despite what it judges to be potentially greater return from other customer groups with need profiles closer to those of general market customers.

Figure 6.5 provides a breakdown of Cray's sales by customer type. "Classical" customers are Cray's original, sophisticated customers who seek the state-of-the-art supercomputer for scientific research. As the internal Cray management report of 14 May 1984 explains, these customers are typically "government users who have purchased primarily on performance. They have demanded very little in the way of software support, IBM compatibility or hardware peripherals. They buy on the basis of speed."[49] "Neoclassical customers (a subgroup of nonclassical customers) are similar to classic customers but tend to be somewhat less sophisticated in that they look to Cray for assistance in developing support software. This group consists largely of oil companies. Finally, the "industrial" customers (the other subgroup of nonclassical customers) are yet more dependent on Cray in that they expect Cray to supply all needed software and software support. These customers are closest to the typical IBM customer who is seeking a data-processing solution, not simply computational speed.

Example A of Figure 6.5 shows that the combined nonclassical customer types

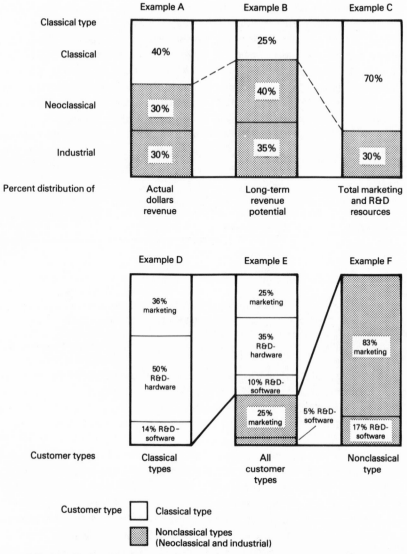

Figure 6.5. Distribution of customer sales and firm resources by customer type. *Source:* The breakdown of customer types by dollar sales and unit sales was provided by Robert Walan, Central Region Sales Manager, Cray Research; the revenue potential figures were provided by Peter Gregory, Vice President of Corporate Planning, Cray Research; the distribution of marketing and R&D invested resources by customer type I derived from data used to develop an instructor's aid. See F. J. Aguiles, "Cray Research, Inc. Teaching Note," No. 5-385-316 (Harvard Business School, October, 1986), p. 2.

contribute 50% more revenue to the business than the classical type (40% versus 60%). More importantly however, as Example B indicates, the long-term revenue potential of the nonclassical customer types is three times as great as the potential long-term revenue contributed by the classical segment. In fact, each nonclassical type *individually* can potentially contribute more revenue than classical customers.

In the absence of competitive considerations, the greater long-term market potential of both the neoclassic and industrial customers, combined or individually, indicates that the firm's resources are best invested in these segments, which together can contribute three times that of Cray's traditional customers. Yet, in fact, as Example C shows, Cray has invested 70% of its R&D and marketing resources in the classical customers. Conversely, only 30% of its marketing and R&D resources are invested in the nonclassical customer types. In summary, Cray has invested 70% of its marketing and R&D in the classical customers while only 30% is invested in the nonclassical types that could potentially contribute three times the revenue of the classical customers. This makes economic sense only when the competitive dynamics discussed earlier, and specifically the dangers of a market-broadening strategy, are considered.

The distribution between R&D and marketing resources within each customer class is also indicative of a commitment to Cray's segment-focused augmentation. Of the firm's resources directed toward the classical customers (Example D) more resources are devoted to R&D in general (64%), as well as *hardware* R&D in particular (50%), than to marketing (36%). This bias is consistent with the nature of Cray's segment-specific augmentation, which unlike IBM's marketing-based general market augmentation, depends entirely on the technical improvement of a core product attribute.

Example F shows that of the total investment in the nonclassical customers, over 80% is in marketing. First note that this amount is no greater than the amount going to marketing for the classical customers (Example E), which has only one-third the revenue potential of the nonclassical customers. More importantly, the fact that over eight-tenths of the nonclassical resources are going to marketing suggests that what effort is being directed at these nonclassical segments does not involve product changes or product line extensions, which could dilute Cray's competitive advantage. Indeed the limited R&D that is directed to the nonclassical types (5% of the total, Example E) is for software and not hardware research and development. Overall, this biased distribution of firm resources toward Cray's traditional customers despite the substantially greater revenue that other customers (closer to the typical IBM customer need profile) could theoretically generate is indicative of Cray's commitment to remaining focused.

ANOTHER FORM OF PRODUCT SEGMENTATION IN THE MAINFRAME MARKET: TANDEM'S FAULT-TOLERANT COMPUTING NICHE

The applicability of the concept of product segmentation is not unique to Cray's behavior in the mainframe industry. Both the customer dimension and the product dimension of product segmentation are readily discernible in the behavior of Tandem Computers. Tandem, like Cray and unlike the BUNCH, serves a small group of customers whose similarity lies in their distinctive, shared product need rather than their industry classification. In particular, a small and otherwise diverse group of customers—airlines, hotel chains, telephone companies, stock exchanges, banks, and a few other financial institutions—are distinguished from the bulk of customers yet similar among themselves in their common need for "fault-tolerant" computing: continuously operating, fail-safe computing with sophisticated networking capability.

Despite the rapid periodic change in various aspects of the hardware technology that comprise computer systems, the fundamental design of computer systems remained essentially unchanged since the origin of the computer until Tandem introduced its first product offering. Up to that point, virtually all computer systems, and all commercially available systems, were based on the previously noted classical, uniprocessor von Newmann architecture, which was so named for John von Newmann, the Hungarian-émigré mathematician who first outlined this approach to computer design in 1946.

The primary design objective of von Newmann architecture is efficiency. Efficiency was of overwhelming importance to early users of information technology because device technology was primitive and specifically because computers were slow and had limited memory capacity. Von Newmann architecture remains today the prevailing design mode because it best meets the needs of the vast bulk of computer system users concerned with the batch processing of data.

IBM's computers, and mainframe computers in general, are designed with the expressed intention of optimizing efficiency in performing batch processing tasks. The efficiency of von Newmann architecture, however, is not without its risk. It is prone to substantial, intermittent failures. While such failures and the resultant "downtime" simply inconvenience batch processing users, they can devastate the interactive, on-line systems indispensable to financial institutions and users with similar needs. Such breakdowns, moreover, can jeopardize the integrity and security of the data used on the systems. When one considers the potential implications of the loss or distortion of data in financial transactions, the urgency of this customer segment's need is apparent.

In May of 1976, Tandem introduced a computer system based on a radically new "multiprocessor" architecture, which in turn, is based on a unique philosophy of design. What has become known as the fault-tolerant or nonstop computer system is designed explicitly to minimize the risk of system failure in on-line transaction processing systems and network environments. More specifically, fault-tolerant computing can be thought of as product augmentation that meets three related customer needs: (1) continuous, around-the-clock operations; (2) immunity to system failure eliminating, for instance, the risk of data loss; and (3) sophisticated networking capability. These needs derive from the special data processing requirements of airline reservation systems, message-handling systems, hotel booking systems, and electronic fund-transfer systems.

The first and most important dimension of product augmentation achieved through this design approach is the fail-safe computer. It virtually never stops running. Nonstop computing in turn reduces the risk of data loss. Finally, the flexible, highly modular design of the system, which also flows from the nature of multiprocessor architecture, is highly effective in meeting sophisticated networking user needs. The company describes its computer system as:

the first general purpose, commercial system designed specifically to fulfill the critical needs of on-line transaction processing. The innovative, fault-tolerant Tandem architecture virtually eliminates the risk of system failures and protects the customers' data bases from damages caused by electronic malfunctions. The system is also the only one on the market that can be expanded modularly—without any programming changes and even while the system is running.[50]

The "glue" that holds the system together is the proprietary, highly sophisticated, complex operating software, "Guardian," that allows multiple processors to be used together ("in tandem") so as to function as a single system. Nonstop operation is facilitated through the software, which in the event that one processor breaks down, responds to modular failure by automatically transferring to another processor the control of the processing being performed. In a similar way, multiple controllers, multiple data-communication paths, and multiple power supplies also assure nonstop operation. A failure in a communication line, for instance, is overcome by using another of several multiple-data paths.

Nonstop system operation, coupled with data-protection features built into Tandem's proprietary data-base-management-system software, "ENSCRIBE," (designed to prevent the damage or destruction of both stored data bases and in-process data) significantly reduce data-loss risk. The protection of data from system malfunction afforded through these features is, to some users, as important as, if not more important than, the nonstop operation feature.

Multiprocessor architecture and Tandem's software also facilitate ease and sophistication in networking. A major criterion for evaluating a computer system has been the number of terminals it can support. Multiprocessor architecture is sufficiently radical to crate the need for a new evaluation criterion: the number of processors that can be linked in one system or in a network. Tandem's architecture permits easy expansion without reprogramming. The basic Tandem system can be expanded from two to sixteen processors to accommodate larger workloads and/or sophisticated networking requirements accordingly. With the introduction of "EXPAND" software in 1979, Tandem made it possible for a single system to use up to 4080 processors in up to 255 different geographical locations.

Unlike the strategies of the BUNCH, Tandem's behavior exemplifies effective product segmentation because it delineated its target segment of otherwise diverse customers on the basis of a product need. Unlike IBM, it has demonstrated an ability and willingness to satisfy this product need. And unlike CDC, Tandem has not deviated from its segment focus. The favorable conditions for product segmentation faced by Tandem—a uniquely augmented product that satisfies an intense need among a *relatively* small but growing group of customers,[51] coupled with the fact of IBM's existing large financial commitment to uniprocessor architecture—has kept Tandem well insulated from general market competition and, in particular, the dangers of head-on competition with IBM.

CHAPTER SUMMARY

This chapter showed that the effectiveness of Cray in the supercomputing niche and Tandem in the fault-tolerant niche was related to their ability to meet the requirements of product segmentation along both the customer dimension and the product dimension. Cray's effectiveness was also seen to be rooted in a clear and sustained commitment to a segment focus despite the lure of a market-broadening strategy.

Conversely, the ineffectiveness of the segment-focused behavior of the other firms was seen to be related to their failure to meet the requirements of product segmentation in the first place or the failure to generate and sustain an adequate commitment to a segment-focused strategy.

Unlike the BUNCH, who failed to define customer segments in terms of product needs, a review of Cray's customer base indicated that Cray was targeting a segment of otherwise diverse customers who were defined by intense, shared, yet distinct product need. A review of Cray's product actions since its founding revealed a strong commitment and a demonstrated ability to implement segment-focused augmentation. On the other hand, the repeated ineffectiveness of IBM's segment-focused strategies was seen to be related to a lack of commitment and ability to implement segment-focused augmentation. Similarly, the ineffectiveness of CDC was seen to be related to its failure to sustain adequate focus and specifically its failure to resist the enticement of greater revenue and reduced strategic risk associated with a market-broadening behavior. CDC's case, in particular, provided a clear example of the generic danger of shifting between, or mixing, a segment-focused strategy with a broad-market-appeal strategy. Such behavior was seen to invite competition at both the high end and low end of a firm's market. This is precisely what happened to CDC, who was squeezed between IBM's superior general purpose computers at one end and Cray's faster supercomputers at the other end.

Evidence of Cray's commitment to its segment-focused augmentation (ultrafast computing) was not only observed in its ongoing record of increasing the limits of computing speed, but also in the demonstrated willingness to sacrifice product reliability, compatibility, and general purpose utility in order to achieve greater speed. Cray's commitment was also observed in its rejection of product-line-broadening projects such as the proposed Quarter Horse computer. Quantifiable evidence of Cray's commitment was observed in the biased distribution of R&D and marketing resources toward Cray's traditional customers, despite the potential for greater revenue from customers with more IBM-like need profiles. Overall, the customer conditions, product actions, and articulated firm intentions reviewed in this chapter show that the concept of product segmentation is a useful heuristic for understanding and diagnosing segment-focused behavior in the mainframe market.

NOTES

1. *U.S. v. IBM*, PX 1043, 1044.

2. J. Rollwagon, quoted in "Cray's Way of Staying Super-Duper," by T. Alexander, *Fortune* (18 March 1985), p. 72.

3. The "BUNCH" is an acronym for Burroughs, Univac, National Cash Register, Control Data, and Honeywell, three of which no longer exist as such due to mergers and acquisitions.

4. See S. T. McClellan, *The Coming Computer Industry Shakeout* (New York, Wiley, 1984) who identifies twenty major categories of segment-focused strategies and even more minor categories.

5. *U.S. v. IBM*, PX 1983.

6. R. Sobel, *IBM v. Japan: The Struggle for the Future* (New York, Stein and Day, 1985), p. 13.

7. M. Magnat, "How to Compete with IBM" (Fortune, 1984), found in *Marketing Management Readings: From Theory to Practice*, Volume III, edited by B. P. Shapiro, R. J. Dolan, J. A. Quelch (Homewood, IL, Irwin, 1985), p. 255.

8. J. Rollwagon, Chairman and Chief Executive of Cray Research, quoted in "Cray Research, Inc.," by F. J. Aguilas, C. Brainard. Available from Harvard Business School Case Services, No. 385-001 (Boston, June, 1986), p. 7.

9. U. Weil, *Information Systems in the 80s: Products, Markets, Vendors*, (Englewood Cliffs, NJ, Prentice-Hall, 1982), p. 129.

10. K. D. Fishman, *The Computer Establishment* (New York, Harper and Row, 1981), pp. 117–118.

11. Quoted from an unpublished written report based on an internal presentation delivered to Cray management on 14 May 1984. Contributors include G. Eastman, D. Harding, K. Hollen, S. Newlin, and S. Noyquist.

12. O. Port, "Super-Fast Computers," *Business Week* (26 August 1985), p. 91.

13. Reinventing the Computer," *Fortune* (5 March 1984), p. 86.

14. Market data on installations and total customers for each year after the introduction of Cray's first computer through 1 January 1985 are presented in Figure 6.3, discussed in a later section.

15. The only other competitor of significance is CDC with 22 percent of the market leaving only 8 percent of the market to be shared among all other supercomputer manufacturers; see "Buble Takes a Bow," *Fortune* (8 July 1985), pp. 9–10.

16. "Reinventing the Computer," p. 86.

17. Weil, *Information Systems in the 80s*, pp. 129, 130, 135.

18. McClellan, *The Coming Computer Industry Shakeout*, p. 287.

19. J. Greenwald, "Frenzied Hunt for the Right Stuff," *Time* (10 August 1987), p. 26.

20. The graininess of experimental parallel processors ranges from only two linked superprocessor nodes to 65,536 very weak processor nodes all linked together and includes a wide distribution of various mid-grain-level machines. For a more technical discussion of the differences in the architectures of experimental parallel processors see D. B. Davis, "Parallel Computers Diverge," in *High Technology* (February 1987), pp. 16–22. Also see P. Hoffman, "The Next Leap in Computer," *The New York Times Magazine* (7 December 1986), pp. 126–133, for a less technical discussion of the state of the art in parallel processing.

21. *U.S. v. IBM*, Dunwell, Tr. 85536–37

22. *U.S. v. IBM*, Dunwell, Tr. 85527; also see Fernbock, Tr. 470–471; E. Block, Tr. 91482, 91678–80.

23. R. Sobel, *IBM: Colossus in Transition* (New York, Truman Talley, 1981), p. 120; Fishman, *The Computer Establishment*, p. 156; DeLamarter, *Big Blue*, p. 53.

24. *U.S. v. IBM*, William Norris, CDC Chairman and Chief Executive Officer, Tr. 5167–18.

25. *U.S. v. IBM*, Gordon Brown, CDC Vice President, Tr. 50996.

26. *U.S. v. IBM*, DX 284, p. 3.

27. *U.S. v. IBM*, Norris, Tr. 5611.

28. *U.S. v. IBM*, Norris, Tr. 5611.

29. *U.S. v. IBM*, PX 93057.

30. *U.S. v. IBM*, PX 1063.

31. *U.S. v. IBM*, Dr. Louis Robinson, IBM Director of Scientific Computing, Tr. 23049.

32. *U.S. v. IBM*, Konoplund, Ex. 90518–20; see also Wright, Ex. 12893–94, 12897.

33. The phrase technological fallout refers to the phenomenon of research and development in one area (i.e., supercomputers) resulting in technological advances that benefit other areas (i.e., general purpose computers).

34. *U.S. v. IBM*, PX 1044.

35. Ibid.

36. *U.S. v. IBM*, PX 1063.

37. *U.S. v. IBM*, PX 1044.

38. *U.S. v. IBM*, PX 1144.

39. Control Data Corporation, Annual Report (1965), p. 5; also see *U.S. v. IBM*, DX 14214.

40. *U.S. v. IBM*, PX 5132.

41. Fishman, *The Computer Establishment*, p. 122.

42. *U.S. v. IBM*, G. Brown, Tr. 50990, 50997; see also DX 13838, pp. 7, 9; DX 13840, p. 8.

43. *U.S. v. IBM*, Norris, Tr. 5618.

44. S. Cray, *Chippewa Falls Herald Telegram* (May, 1972).

45. J. Rollwagon, quoted in Alexander, "Cray's Way of Staying Super-Duper," p. 76.

46. Alexander, "Cray's Way of Staying Super-Duper," p. 76.

47. J. Rollwagon, quoted in Alexander, "Cray's Way of Staying Super-Duper," p. 72.

48. Weil, *Information Systems in the 80s*, pp. 130, 135.

49. See footnote 12.

50. *U.S. v. IBM*, DX 13945, p. 15.

51. The investment bank, Mogran Stanley, forecasts growth of 30 to 40 percent annually for the small group of customers requiring continuously operating, fail-safe data processing with elaborate networking; see Weil, *Information Systems in the 80s*, p. 141.

7

Conclusion

CONCLUDING SUMMARY

This book has specified the essential features of behavior in heterogeneous markets (Chapters 2 and 3) and has showed the applicability of the concepts of behavior developed to both emergent and intended strategy in the market for computer mainframe products (Chapters 4, 5, and 6). A new and fundamental idea of this study is the dissection, categorization, and organization of market behavior based on the concept of "two-dimensional market heterogeneity." More specifically, the idea of two-dimensional market heterogeneity is the basis for delineating basic types of market behavior, for developing a more balanced view of the nature of behavior and the function of markets, and for identifying demand-based sources of market effectiveness and superior business performance.

Each category of market behavior delineated meshes the demand logic of economic analysis with the customer-effectiveness perspective of marketing in the development of new constructs of behavior useful in the field of business strategy. This study has introduced the business strategy concept of "product augmentation," which is based on linking the economic logic of product differentiation with the augmented product concept of marketing (Chapter 2, the section "Product Heterogeneity and Product Dimensional Behavior"). Another type of behavior delineated is "price segmentation," which is rooted in the economic logic of price discrimination but extended to include behavior that does not entail the use of explicitly discriminatory pricing (Chapter 2, "Customer Heterogeneity and Customer Dimensional Behavior"). This book has also developed the new construct of "product segmentation," which very roughly resembles the broader marketing idea of market segmentation but is here given a fuller logic, and a more precise definition, in that it incorporates in one concept elements of the logic of both product differentiation and customer disaggregation (Chapter 2, "Two-Dimensional Market Heterogeneity and Two-Dimensional Market Behavior").

In addition to providing a conceptual foundation for the development of these behavior categories, two-dimensional market heterogeneity leads to a more balanced perspective on the function of markets, the objective of that function, and the nature of the mechanism facilitating that function. More specifically, two-dimensional market heterogeneity implies a fuller view of markets to include their function of "matching" heterogeneous products with heterogeneous customers toward the end of not only productive efficiency, but also customer effectiveness, and through, not just the price mechanism, but the superordinate vehicle of market information. The emergence of the importance of information leads to the identification of another type of market behavior: the strategic use of information inadequacy and its effects (Chapter 3, "Information Inadequacy"). Contributing to understanding these effects are the theoretical economic ideas of asymmetric information, information-acquisition costs, and returns to reputation, as well as the marketing ideas on information signaling, low buyer sophistication, and perceived customer purchase risk.

The concepts developed are useful not only in describing actual market behavior but also in identifying basic sources of market effectiveness and therefore in diagnosing market strategy. Market effectiveness is related to the strategic use of two-dimensional market heterogeneity and its implication for information. More precisely, sources of market effectiveness include product augmentation, price segmentation, product segmentation, and the exploitation of the effects of inadequate information.

The ideas of product augmentation and the strategic use of the effects of information inadequacy are especially useful in assessing the sources of IBM's customer appeal (Chapter 4). The concept of product augmentation as developed in this thesis combines the distilled economic logic of product differentiation with the augmented product concept of marketing. This hybrid concept provides a more robust understanding of IBM's behavior than either the economic or marketing concepts do separately. While the economic logic component of product augmentation is important to firm conduct generally, it is unusually critical to understanding specifically IBM's behavior due to the context in which it is immersed. In particular, the nature of products, customers, and competition in the mainframe industry coupled with IBM's position in that industry intensify the danger and diminish the effectiveness of price-competitive behavior for IBM, while they create conditions conducive to the effective use of risk-reducing product augmentation.

The perspective of the augmented-product concept, originating in marketing literature, is especially important to understanding IBM's behavior because of the less tangible nature of IBM's augmentation. The augmented-product concept rejects the economic notion of a product as a commodity; it requires that it be understood in terms of its widest potential for providing customer benefits. Consequently, the presence of "nonhardware" differences (between IBM and non-IBM products) that remain after "product differences are removed" are now viewed as part of those product differences and, in fact, are shown to be the basis of IBM's product augmentation strategy.

The idea that product augmentation along non–price/performance dimensions, and specifically along less tangible risk-reducing dimensions, is an important factor in IBM's customer appeal received empirical support through a variety of observations. Empirical economic research, customer survey data, IBM's own internal product assessments, as well as testimony by IBM executives and industry observations by others all suggest that, even in the heyday of its technological innovativeness, IBM's advantage could not be explained by superior price/performance.

The evidence pointed toward IBM's image of offering more in terms of "nonhardware services" as a source of its market effectiveness.

This idea is further supported with detailed statistical data on purchasing behavior. The data indicate that customers in general purchase mainframe products because of customer support, product reliability, and firm stability—factors that appeal to the effects of information inadequacy and, specifically, low buyer sophistication, high receptivity to information signaling, and perceived customer purchase risk. Conversely, price-related factors are seen to be among the least significant in determining purchasing behavior.

The data established the important empirical finding that those who select IBM are also those whose purchasing situation (greater general management influence and participation), purchasing behavior (larger buys among larger customers entailing fewer pilots), purchasing preferences (greater importance of support, lesser importance of price), and actual ratings of risk indicate a greater need for reduced purchase risk. The fact that these customers with higher perceived purchase risk are also those customers who select IBM, in turn, indicates that product-dimensional, risk-reducing behavior is an important source of customer effectiveness for IBM. IBM's strategy is operationalized in the marketing techniques of bundling and account management, as well as the extensive use of non–sales field personnel, which either improves the perception IBM's product as the low-risk choice or intensifies low-risk purchasing behavior thereby facilitating risk-reducing product augmentation.

The concept of price segmentation is also helpful in discerning the strategy in IBM's behavior (Chapter 5). In part due to the existence of, and IBM's encouragement of, the phenomenon of customer lock in, IBM is able to extract higher-than-normal levels of revenue and profit from its established, less price-sensitive customers while limiting competition among the newer, more price-sensitive customers through unprofitably low pricing (subsidized by established customers). More generally, observation of actual behavior reveals the applicability of generic analytic frameworks of price segmentation. Specifically, IBM's tying card sales to mainframes, metering of usage through extra-use chargers, and practice of bundling processors with memory are all revealed to be actual cases of the theoretical categories of price segmentation (discussed in Chapter 2). Observations also reveal other forms of price segmentation not explicitly addressed in the literature: "functional pricing" of submodels, educational discounting, "stable charges" across product generations, interface manipulation, and the variation of bundles through the disproportionate allocation of field support.

The presence of this segmented pricing in IBM's behavior is hard to discern because in all these cases, with the exception of functional pricing and educational discounting, price differences are not explicit. In the case of functional pricing, model-level profit data hides the pattern at the submodel level. The characterization of educational discounting as a charitable action benefiting nonprofit institutions; the idea that maintenance and/or accelerated depreciation fully explain higher extra-use charges; the use of constant rate apportionment methodology; and the tying together, bundling, or repackaging of products all serve to camouflage the presence of segmented pricing. Irrespective of the legality of IBM's behavior, the evidence presented in Chapter 5 shows that such pricing is based not on cost but rather on the strategic objective of protecting or expanding sales among more price-sensitive customers and extracting maximum revenue from less price-sensitive customers.

The new concept of product segmentation is also useful in diagnosing behavior in the mainframe industry (Chapter 6). By integrating the idea of customer segmentation with the economic logic of product differentiation and its consequent implications for insulating a firm from competition, a concept of behavior emerges that is more useful in diagnosing segment-focused strategy in the mainframe industry than the broader marketing concept of market segmentation.

A review of segment-focused behavior suggests that the effectiveness of Cray in the supercomputing niche, as well as Tandem in the fault-tolerant niche, is related to their ability to meet the requirements of product segmentation as to both customer segmentation and product augmentation. Observed behavior also suggests that Cray's effectiveness is rooted in its clear and sustained commitment to a segment focus despite the lure of a market-broadening strategy. Conversely, the ineffectiveness of the segment-focused behavior of the other firms is seen to be related to their failure to meet the requirements of product segmentation in the first place or the failure to generate and sustain an adequate commitment to a segment-focused strategy.

Unlike the BUNCH, who failed to define customer segments in terms of product needs, a review of Cray's customer base indicates that Cray is targeting a segment of otherwise diverse customers who are defined by an intense, atypical, shared product need. A review of Cray's product actions since its founding reveals a strong commitment and a demonstrated ability to implement segment-focused augmentation. On the other hand, the repeated ineffectiveness of IBM's segment-focused strategies is seen to be related to a lack of commitment and ability in implementing segment-focused augmentation. Similarly, the ineffectiveness of CDC is seen to be related to its failure to sustain adequate focus and specifically its failure to resist the lures of greater revenue and reduced strategic risk associated with market-broadening behavior.

Evidence of Cray's commitment to a segment-focused augmentation is not only observed in its ongoing record of increasing the limits of computing speed but also in its demonstrated willingness to sacrifice product reliability, compatibility, and general purpose utility in order to achieve greater speed. Cray's commitment is also observed in its rejection of product-line-broadening projects, such as the proposed Quarter Horse computer. Quantifiable evidence of Cray's commitment is observed in the biased distribution of R&D and marketing resources toward Cray's traditional customers, despite the potential for three times higher revenue from customers with more IBM-like need profiles.

Overall, the observation of customer purchasing behavior (Chapter 4), pricing actions (Chapter 5), customer conditions (Chapter 6), product actions (Chapters 4, 5, and 6), and firm intentions as explicitly articulated by firm members (Chapters 4, 5, and 6) indicate that product augmentation, price segmentation, product segmentation, and the exploitation of the effects of information inadequacy are useful heuristics for understanding strategy and its effectiveness in the market for mainframe computer products.

DYNAMICS BETWEEN AND WITHIN STRATEGY TYPES

Much work still needs to be done in understanding the relationship among the strategy types and the types of substrategies within each major strategy category discussed above. However, patterns in actual behavior in the mainframe products industry seem consistent with initial theorizing in Chapter 3 as to the relationships among, and substrategies within, the three categories of heterogeneous behavior. First, observed behavior seems consistent with theorizing as to complementarity and potential incompatibility between strategy types. Second, the theoretical distinction between the alteration of the perceived product versus the alteration of customer wants as alternative approaches within each of the three strategy types is useful for understanding different IBM marketing techniques.

As to the relation among strategy types, and complementarity in particular, IBM's behavior suggests that both product augmentation and price segmentation are not incompatible and may be mutually reinforcing. IBM's policy of offering support to customers according to their needs is a vehicle to both product augmentation and price segmentation. product augmentation is facilitated in that the customer is assured of receiving the support he needs thereby contributing to the perception of IBM's product offering as the low-risk choice (Chapter 4). The same policy facilitates price segmentation in that the consequent variation in levels of support received by customers is a means of extracting consumer surplus (Chapter 5).

At a more detailed level, product bundling, as practiced by IBM, serves as both a means to product augmentation and price segmentation. It facilitates product augmentation not only by providing unspecified support but also by reducing the complexity of the purchase choice faced by unsophisticated buyers (Chapter 4). It facilitates price segmentation not only through varied support levels but also through the phenomenon "preference reversal," as shown to apply to the mainframe customers (Chapter 5). The facts that one marketing policy (support according to need) and one specific marketing technique (bundling) achieve the objectives of both product augmentation and price segmentation render moot the notion of conflict between the strategy types due to the need to allocate scarce organization resources. Resource-allocation-based conflict is minimal because one policy or activity, and thus one set of costs, is shared by two strategies.

Not only do these two strategy types not necessarily conflict, but they are also, to some degree, mutually reinforcing. The band of price insensitivity resulting from IBM's effective product augmentation in the areas of service, support, and reliability creates conditions conducive to segmented pricing in that it reduces product substitutability. Conversely, the extra high prices associated with IBM's high-end machines, although symptomatic of IBM's segmented-pricing program, serve to reinforce the image of IBM as a high-quality computer manufacturer insofar as price signals product quality. Hence IBM's product augmentation reinforces its price segmentation and its price segmentation reinforces its product augmentation.

Another theoretical relationship between the strategy types discussed in Chapter 3 seems consistent with observed behavior. Specifically, the record of emergent strategy in the supercomputing niche prior to Cray seems consistent with the idea of incompatibility in mixing a segment-focused strategy with a broad market strategy

(Chapter 6). Compatibility problems are suggested by IBM's failed attempts at mixing a strategy focused on the ultrafast-computing customer segment with an ongoing, broad market strategy directed at the much larger base of general-purpose computing customers. This is especially apparent in IBM's second failure in which it was unable to mix merely a sustained "presence" (nonetheless a profitable presence) in the supercomputer niche with its mainstream general market strategy. Even more suggestive of incompatibility is the record of CDC's strategic behavior, which provides a case study on the generic danger of mixing or shifting between a segment-focused strategy and a broad market strategy. Such behavior was seen to invite competition in the customer segment and in the general market, with the result that CDC was squeezed between the superior general-market competitor (IBM) and the superior segment-focused competitor (Cray).

If all the hypothesized tradeoffs and complementarities discussed in Chapter 3 are assumed to be correct, then among the possible strategies and strategy mixes (listed in Table 3.2), the only three that emerge as not internally compatible—price competition alone, product segmentation alone, and product augmentation/price segmentation together—are the same three that describe the patterns of behavior observed in the mainframe computer industry (discussed in Chapters 4, 5, and 6 respectively).

Observed behavior also seems to be consistent with theorizing about substrategies within each major strategy type. In particular, observed behavior seems consistent with the idea that each strategy type has two sides: the alteration of the perceived product in order to more closely match customer wants and the alteration of customer wants to more closely match an existing product advantage. This theoretical distinction is directly applicable to both the product dimensional behavior of IBM (product augmentation) and the customer dimensional behavior of IBM (price segmentation).

IBM implements product augmentation through two different marketing techniques, each of which alters a different side of the product/customer matching equation. Product bundling alters the nature of the perceived product to better match the customer's desire for a low-risk purchase, while account management intensifies the customer's proclivity toward a low-risk purchase by exaggerating the role of general management in the buying process. This, in turn, intensifies risk-averse behavior not only by expanding the proportion of inexperienced buyers in the purchase process but also by increasing the personal risk of lower-level expert buyers as a result of greater upper-management scrutiny. In one case, IBM is altering the product, in the other, IBM is altering customer wants; but the price-sensitivity-reducing effect of a closer matching of products with customers is achieved in both cases.

Similarly, the effectiveness of IBM's price-segmentation strategy lies not only in the manipulation of product characteristics (in this case the price of the product) but also in the manipulation of customer characteristics (in this case customer sensitivities as to price). In addition to exploiting existing differences in customer price sensitivity through differential pricing, IBM can manipulate customer price sensitivities through any actions that create data processing system incompatibilities in general, and through the manipulation of hardware interfaces in specific. To the extent that IBM not only exploits but creates the lock-in of a group of its customers either within its customer base or between its customer base and the total market, IBM is not simply using product pricing to achieve the goal of price segmentation but is also altering customer characteristics toward the end of segmented pricing. On the customer side of the implementation of segmented pricing, IBM is both creating the conditions for the effective use of tiered pricing in the first place and intensifying price insensitivity among less price-sensitive customers, thereby amplifying consumer surplus extracted. As is the case with its product-augmentation strategy, IBM's price-segmentation strategy is implemented through the alteration of both the product characteristics and the customer characteristics.

SIGNIFICANCE FOR IDEAS ON WELFARE

This book is concerned with market behavior. It is not about what, in IO economics, is termed "industrial performance." More precisely, this thesis is not about welfare in any of its definitions, government policy attempting to promote welfare, or the particular welfare policy questions specifically concerning IBM's behavior. Nevertheless, the ideas on behavior developed here, and the perspective they lead to, have significance for ideas on welfare, public policy, and specific competing viewpoints on IBM's behavior.

Conventional wisdom correctly asserts that product differentiation and price discrimination interfere with the price mechanism, which facilitates the most efficient production of goods resulting in the lowest prices to the consumer. Product differentiation leads to "monopoly power over price" and, consequently, higher prices to consumers. In price discrimination the seller extracts the *"consumer's* surplus."[1] Indeed, any behavior that interferes with the price mechanism is seen as detrimental to welfare because it diminishes efficiency. Such behaviors are systematically discouraged as a matter of public policy, which is formalized in competition legislation. Conversely, the type of behavior typically promoted as a matter of policy and legislation is that which assures the proper functioning of the price mechanism and thus optimal efficiency: price competition.

The significance of two-dimensional heterogeneity for ideas on welfare and for public policy lies in the potential incompatibility it creates for the goals of efficiency and effectiveness. Given the assumption of homogeneity, differentiating factor, the efficient production of goods is consistent with effectiveness in customer satisfaction. Low production costs mean low prices, which given homogeneity, are the sole product benefit important in customer satisfaction. Under homogeneous conditions, therefore, price competition is the primary path to customer welfare. However, the fact that price competition, given homogeneous markets, may be ideal does not necessarily imply that price competition per se is ideal (or that homogeneity is to be preferred over heterogeneity).

Two-dimensional heterogeneity creates potential incompatibilities in the goal of efficiency and effectiveness because price and net product benefits are no longer necessarily one and the same. As a result, the path to welfare is not as simple. Different types of behavior may simultaneously improve effectiveness yet reduce efficiency or vice versa. The goal of producing products at the lowest cost and providing them to customers at the lowest price must now be balanced against the need to provide customers with new and different products to better match products with heterogeneous customer wants. All four sources of effectiveness that arise from two-dimensional heterogeneity (product augmentation, price segmentation, product segmentation, and the strategic use of information inadequacy), therefore, have significance for ideas on welfare and public policy rooted in welfare ideas.

The significance of product-dimensional behavior (both product augmentation and product segmentation) for welfare and public policy lies in its effects on customer choice and market growth. More specifically, product-dimensional behavior enriches customer choice and creates incentive for market growth. First, product-dimensional behavior expands product diversity thereby extending the range of consumer choice,[2] a condition for the better matching of products with heterogeneous customers. Second, product-dimensional behavior can motivate growth. The lures of "extranormal" profitability and a superior market position, which can result from general-market augmentation or segment-specific augmentation, motivates investment in such product augmentation efforts (efforts with risks that might otherwise inhibit this investment).

Consistent with the distinction between two dies to the implementation of each strategy type, product-dimensional behavior stimulates growth not only from the seller side through the enticement of extranormal profitability but also from the buyer side. The demand for a given type of homogeneous good is limited. At some point customer demand is satiated and product growth abates. Product augmentation can create a range of new product uses and applications to customer problems for which previous product solutions were unavailable. The creativity of the buyer in applying new product augmentations can stimulate customer demand beyond that in a homogeneous product market. Product dimensional behavior therefore can contribute to market growth by both motivating the seller and expanding the demand of the customer.

Just as product augmentation and product segmentation can have a positive impact on welfare through greater effectiveness in matching products and customers, so too can price segmentation improve welfare. Price segmentation can be viewed as one type of enhanced seller sensitivity to customer differences. Similarly, it can be viewed as a manifestation of greater sensitivity to differences in what customers can afford. Scherer argues that "through such discriminatory pricing [sellers can] respond sensitively to consumer's variegated wants, and at least from the standpoint of allocative efficiency, their responsiveness is laudable."[3] More generally, Spence shows that appropriate forms of price discrimination can result in a more "desirable range of price-quality choices" so that from a welfare perspective consumers are faced with a more nearly "optimal product variety."[4]

Some spatial economic analysis suggests that price discrimination can lead to lower prices and higher output than nondiscriminatory pricing. Norman finds that increasingly competitive spatial models—from a multiplant monopoly, to Loschian competition, to the even more competitive Greenhut–Ohta model—give price-output solutions that on the surface are consistent with spaceless neoclassical microeconomic expectations (higher output and lower prices) but in fact are achieved through greater, not lesser, price discrimination.[5] Hence price discrimination is associated with increased not decreased competition, or[6] more strongly, "price discrimination acts as a check upon local market power."[7] Dorward summarizes the welfare implications:

In the spatial economy, whenever the behavioural assumptions do not restrict the firm to viewing its market as a fixed area within which it attempts monopoly pricing, as with multi-plant monopoly or Loschian competition, the increased price discrimination consequent upon increased competition will give progressively higher levels of welfare per unit area than f.o.b. pricing.[8]

Segmented pricing can increase the availability and extend the provision of goods or services to customers who would otherwise be unable to purchase these at the single market price resulting from price competition. Similarly, price segmentation can attract demand to underutilized capacity and resources. Insofar as the greater provision of such goods and services, or the greater utilization of such capacity and resources, is beneficial to customers, price segmentation enhances welfare. Hence, roughly similar to the effects of product-dimensional behavior, price segmentation can stimulate demand and can increase the range of products that a customer can afford to purchase. Therefore, heterogeneous market behavior overall, and specifically product augmentation, product segmentation, and price segmentation, can contribute to welfare by expanding the range and/or the volume of products purchased by customers through the stimulation of the buyer and/or the motivation of the seller.

It is not clear that the net impact of the positive and negative effects on welfare from heterogeneous behavior, and specifically the net impact of enhanced customer effectiveness and diminished efficiency, represent a lower state of welfare than that resulting form price competition. This statement should not be misunderstood as reflecting the ideas of the theory of "second best." The theory of second best makes the important identification of a flaw in the reasoning that "if we cannot attain the ideal state of competition, the next best alternative is to approximate it as closely as possible."[9] Second-best theory disputes this logic; it argues: "A situation in which more but not all of the optimum conditions are fulfilled, may *not* be superior to a situation in which fewer are fulfilled; indeed, we cannot state *a priori* which case is likely to be superior."[10]

This thesis, however, suggests something more than this. It is not challenging the superior desirability of the next closest approximation to ideal price competition as second best; it is suggesting that the so-called first-best pure price competition is not necessarily superior to other behaviors *in two-dimensional heterogeneous markets*.

Once the potential incompatibility of efficiency and effectiveness that arises from two-dimensional market heterogeneity is recognized, pure price competition is no longer necessarily "ideal" or "optimal" as far as consumer welfare goes. Once this more complex concept of welfare containing potential internal incompatibilities is adopted, other forms of market behavior emerge as indispensable components in a range of behaviors that contribute to welfare.

Unlike homogeneous markets in which price-competitive behavior is the exclusive and internally consistent path to welfare, in heterogeneous markets all the

behavior types can both enhance and diminish welfare. In two-dimensional heterogeneous markets, more price competition can enhance welfare through its impact on
efficiency and diminish welfare through its negative impact on effectiveness. Similarly, the three heterogeneous behavior types can both enhance and diminish welfare
through their influence on efficiency and effectiveness but in a reverse manner.
More explicitly, heterogeneous behavior can enhance welfare through its impact on
effectiveness and can diminish welfare through its impact on efficiency.

The net impact of these multiple behavior types in heterogeneous markets is not
clearly necessarily less desirable than the sole use of price competition in what
would thus become a homogeneous market. To view price competition as the ideal
case is to deny the significance of two-dimensional market heterogeneity for welfare
and specifically the dependency of welfare on the potentially conflicting subgoals of
efficiency and effectiveness. The significance of two-dimensional market heterogeneity for welfare suggests an avenue for further research, beyond the theory of
second best, that investigates the very premise of price competition as the first best
and considers the possibility of "alternative bests."

The welfare implications of two-dimensional market heterogeneity further suggest an avenue for *applied* research. In particular, the net impact on efficiency and
effectiveness of heterogeneous behavior in the mainframe computer industry merits
further investigation. Such research might weigh negative welfare effects against the
benefits of IBM's provision of unspecified support and Cray's and Tandem's segment-specific augmentations that widen customer choice and facilitate the more
precise satisfaction of customer wants.

SIGNIFICANCE FOR PUBLIC POLICY

The idea that price competition is an ideal is an essential feature of what has been
called the "structuralist paradigm." Block summarizes the features of this structuralism as currently conceived:

- The perfectly competitive model is the ideal.
- The real-world markets are usually seen as a continuum from monopoly to perfect
  competition.
- Monopoly power exists in greater or lesser degree in any market that is not
  perfectly competitive.
- Excessive profits are unfair and constitute a misallocation of resources.[11]

Government policies promoting, discouraging, and generally regulating various
types of market behavior are rooted in the structuralist paradigm as presently formulated. Such public stances toward business are formalized in competition legislation
or what in the United States are poorly labeled "antitrust" laws. The fact that this
legislation is rooted in the form of structuralism outline above is readily acknowledged by those responsible for such legislation.

The U.S. Justice Department, for instance, reported that the overwhelming consensus of its officials is that "the extent and degree of competition in any given
industry is judged primarily on the basis of the structural characteristics of that

industry" and on the presence of *"vigorous price competition"*[12] (my emphasis). Similarly, the U.S. Supreme Court reports that the intent of the law is to be "in conformity with the simplest [structuralist] models."[13] Stern and Grabner point out that this government posture grows directly out of structuralist economic ideas that view price competition as an ideal: "The viewpoint of the federal government is, in general, the same as that of industrial organization economists. The primary tool employed to analyse and evaluate competition in any given industry is the price theory which arose from the norm of perfect competition."[14] Block argues that the narrow view of competitive behavior in structuralism as conceived in neoclassical economic thought has influenced the court's approach to competition:

Coincident with the development of price theory models of the neoclassical structuralist economists . . . competition lost virtually all of its behavioural meaning. . . . Because of this unfortunate theoretical development, economists virtually cut themselves off from a study of competitive behaviour in a real world setting. . . . While the courts could hardly ignore behaviour, they were enough influenced by neoclassical economic thought to ignore all behaviour except that explicitly linked to price. . . . "Perfect competition" became the standard by which other markets were judged.[15]

The idea of price competition as a "standard" or "norm" has naturally led to a public policy posture that promotes price competition and discourages certain other forms of product- and customer-dimensional behavior as "anticompetitive." However, to the extent that such policy fails to recognize the significance of two-dimensional market heterogeneity for welfare and, in particular, fails to appreciate the importance of effectiveness in welfare and its potential conflict with the traditional goal of efficiency, it may be ineffective in promoting welfare if not positively counterproductive. The significance of two-dimensional market heterogeneity for public policy lies in the recognition that the promotion of welfare does not necessarily imply the promotion of price competition (or the discouragement of certain forms of heterogeneous behavior). Where conventional wisdom views price competition as the barometer of competitiveness and the ideal for purposes of welfare, it is here seen as just one of several forms of behavior that does not necessarily represent a more vigorous state of competitiveness or a higher level of welfare than other forms of market behavior. Conversely, certain types of heterogeneous behavior that conventional wisdom may discourage are here seen to not necessarily represent either less vigorous competitiveness or a lower level of welfare.

Thus, for example, the spatial economics literature that links greater price discrimination with greater levels of competition and greater welfare has important implications for antiprice discrimination policy associated with, for instance, the U.K. Price Commission (prior to its abolition) and the U.S. Robinson-Patman (antitrust) Act. As Capozza and Van Order explain: "If firms selling in a series of separated markets adopt the non-discriminatory pricing policy to which legislators appear to be committed, competitive entry may well lead to a general increase in prices."[16] Many spatial economists concur with Greenhut, Nopman, and Hung, who conclude: "It is necessary . . . that the methodological foundations of this legislation be examined: it will emerge from this examination that much of the legislation should be reevaluated."[17] Dorward points out that spatial an analysis "implies the need for a rather complex form of price regulation. Indeed, it could be the case that the costs of implementation would exceed the benefits."[18]

In fact, the promotion of price competition (and/or the discouragement of other forms of behavior) not only may be ineffective in achieving welfare, but may work against welfare. Public policy encouraging the ideal of price competition may create disincentives to achieving the very ends toward which the policy is ultimately directed. Because the successful promotion of price competition leads to "normal profitability," proprice competition policy can reduce welfare insofar as it removes the incentive for product diversity, market growth, and the availability of products to those who can least afford them.

The more complex idea of welfare that two-dimensional market heterogeneity implies in turn suggests a more complex role for public policy. Policy should not be directed toward promoting price competition per se but should be more directly orientated to the ultimate goal of welfare, which may or may not imply proprice competitive policy. Specifically, policy should consider the need to balance the promotion of price competition against the need to encourage investment by firms in disequilibrating heterogeneous behavior that leads to both abnormal profits and the better matching of products with customer wants.

The implication of two-dimensional market heterogeneity for information and, in particular, the importance of information to both the buyer in evaluating products and to the seller in understanding product benefits needed points to a further new role for public policy: the promotion of information availability and the reduction of information costs.

Consistent with the distinction between buyer- and seller-side approaches to heterogeneous behavior, the significance of information for welfare is two-sided. On the seller side, information assures that the seller knows the types and areas of product augmentation and customer segmentation, or both, that are needed. On the buyer side, information assures that purchasing behavior accurately directs the incentive for product augmentation and customer segmentation, or both, toward areas of consumer preference.

SIGNIFICANCE FOR SPECIFIC, COMPETING VIEWS ON IBM'S BEHAVIOR

The generic tendency to categorize behavior in terms of a continuum between the endpoints of price competition and monopoly and to assess competitiveness in terms of the degree of price competition has led to unnecessarily narrow and dichotomous views of IBM's behavior. On the one hand, IBM's behavior is characterized as "anticompetitive," "abusive," and as an exhibition of "monopoly power." Among the most articulate works representing this viewpoint is that of Brock in the area of IO economics[19] and, more recently, DeLamarter in the area of business history.[20] On the other hand, IBM's behavior is portrayed in terms of price/performance competition and characterized as "vigorously competitive." Among the best spokesmen for this view are Fisher, McGowen, and Greenwood[21] in business studies and Fisher, McKie, and Mancke in the area of economic history.[22]

Because both specific viewpoints are rooted in a perspective that sees competitiveness in terms of the level of price competition, each specific viewpoint shares the same emphasis on the importance of price premiums (or lack thereof), which are indicators of the degree of price competition. In the first view, IBM's price premiums exist due to abusive, anticompetitive monopoly behavior and market share dominance that both results from and gives rise to such behavior. According to the second view, price premiums are actually an "illusion" created by the superior physical performance of IBM hardware and, in particular, resulting from the rapid improvements in computer performance due to technological innovation by IBM. Unlike the first view, which sees extranormal profits as the result of unfair, exploitive behavior, the second view sees such profits as the just reward for superior hardware performance resulting from ongoing innovativeness. In short, one view attempts to reduce IBM's behavior to price behavior to show it is competitive, and the other view attempts to establish a lack of price competition to show diminished competitiveness.

It is not the object of this study to contest or advocate legal opinions, to resolve public policy debate over the degree of IBM's "competitiveness," or to assess the desirability of government-imposed constraints on its behavior and changes in its structure. Nor is it the goal of this study to discredit empirically the competing viewpoints of applied economists or to provide a review of the evidence amassed purporting to support one side or the other. This study does not argue that one view is more correct than another; indeed both contain elements of truth.

However, this study *is* significant for these specific views of IBM's behavior because it implies that both views are inadequate for the same reason. More specifically, each view fails to portray the full dimensionality of IBM's behavior. This is not to say that IBM's behavior is more appropriately represented as some combination of, or midpoint between, the poles of the unidimensional price competition/monopoly continuum, but that the continuum itself is inadequate in portraying the heterogeneous types of behavior revealed in Chapters 4, 5, and 6.

The need to judge IBM's competitiveness in terms of price competitiveness implicit in both views is symptomatic of an underlying perspective that fails to reconcile the possibility of vigorous competition (in the form of heterogeneous behavior) with a lack of price competitiveness. While the second view portrays IBM's behavior as competitive but fails to recognize the full nature of that competitiveness, the first view misconstrues a lack of price competitiveness as necessarily implying a lack of competitiveness in general. Whatever the degree of IBM's competitiveness, it should not be judged solely in terms of the absence or presence of price competitiveness.

For the purposes of scholarly inquiry such a mindset promotes a biased inquiry into the nature of computer industry behavior and results in a disproportionate emphasis on only two types of behavior to the neglect of other behavioral types. Moreover, although legal opinions and scholarly viewpoints remain distinguishable, it seems the legal debate surrounding IBM's behavior has blurred this distinction, and legal positions have crept into scholarly representations of IBM's behavior with the effect that the unidimensional and bipolar nature of these academic viewpoints has been more emphasized than would otherwise be the case.

The inability to explain price premiums or extranormal profit in terms of performance differences does not necessarily imply a reduced level of competition and the presence of monopoly power. On the other hand, competitiveness need not imply the presence of price/performance competitiveness and therefore the performance superiority of IBM's hardware. The proclivity to account for IBM's success or, more narrowly, the presence of price premiums solely in terms of either superior product performance or the anticompetitive activities of a monopolist is symptomatic of a mindset that fails to recognize the full dimensions of competitiveness and sources of competitive advantage. A more balanced perspective on IBM's competitiveness and sources of competitive advantage requires consideration of risk-reducing heterogeneous behavior in addition to price/performance competition and the abuse of monopoly power.

The myopia of the second viewpoint (advocating the price/performance competitiveness of IBM) is more subtle than that of the monopoly power viewpoint because it draws on the role of innovation in making its argument for vigorous price competition. However, in dynamic markets characterized by rapidly changing and complex technology and, consequently, inadequate information, the ability of the customer to readily compare products on the basis of price is diminished. Thus the very feature that the modified price-competition view invokes in arguing for the presence of price competition—technological innovation—is the same feature that, in fact, tends to reduce the ability to compete on price.

The significance of innovation in interpreting behavior in the mainframe industry is not that innovation in some way hides the level of price competition, but that it points to the importance of heterogeneous forms of competitive behavior. Indeed, the insistence by the expert witness economist, Fisher, to reduce innovative behavior to price behavior may be symptomatic of a more general disciplinary tendency. As Armstrong laments:

Despite these frequent reminders and the simple common sense of the Schumpeterian observations, static, innovationless price theory has remained the stock trade of most economists. As Robert Solow has complained: "The multifaceted dynamics of technological advance and industrial transformation . . . are almost wholly excluded from the view of Establishment economics. The notion that a competitive price-directed market is the underlying economic reality lingers on, a fixation even of those who proclaim the organization revolution.[23]

There is, further, no necessary reason why the dominant firm should be the leader in innovation as Fisher implies. Early arguments focused on the greater resources of larger firms as providing a basis for leadership in innovation. Recently, however, more credibility has been given to the recognition that the cultivation of an organizational culture conducive to innovation may be more easily established in smaller firms with more flexibility to adopt the latest technological developments in both products and manufacturing processes.

Brock provides empirical evidence that IBM has, in fact, inhibited innovation,[24] thus pointing to the ironic possibility that the factor of technological change is not what gives IBM a competitive advantage (as the price/performance view argues), but in fact may be what prevents IBM from establishing a monopoly. Put positively, innovation is what makes the industry competitive, and strategic behavior so important, *despite* the concentrated industrial structure and, in particular, IBM's dominant market share. In this alternative view, technological followership is not a form of strategic weakness[25] but rather an important component of a more general posture of "adaptive strategy"[26] and a key to continued market dominance.

The role of innovation in sustaining a state of market disequilibrium leads to the identification of an anomaly in the underlying paradigm in which the specific, competing views are rooted. The nature of technological innovation in mainframe markets and, in particular, the complexity and high rate of change of technology contribute to market conditions that inhibit both price competition and the establishment of monopoly power. In other words, the normal tendency toward price competition in markets is constrained, yet at the same time, the potential for establishing monopoly power is limited by the destabilizing effect of rapid innovation. Innovation inhibits price competition on the one hand yet inhibits monopoly power and encourages competitive behavior on the other.

The simultaneous inhibition of price competition and encouragement of competitiveness represent an anomaly[27] for a framework that views competitiveness in terms of price competition. The underlying framework of the competition policy debate regarding IBM's behavior does not easily reconcile the lack of price competition with a state of vigorously competitive market behavior. Such a phenomenon, however, is not anomalous in the context of the fuller viewpoint presented in this thesis. In other words, what in the existing structuralist paradigm appears to be an anomaly—a state of vigorous competition despite a lack of price competition—when considered in the context of the full anatomy of competitive market behavior developed here is revealed to be not an anomaly at all. In fact, what appears to be an anomalous phenomenon in the structuralist paradigm, is, arguably, the prevailing condition of mainframe market behavior.

SIGNIFICANCE FOR BUSINESS POLICY

The perspective of two-dimensional market heterogeneity has significance for business policy research and practice. Its research significance lies in the potential wider applicability of the framework for thinking about the demand side of strategy and the deeper development of the concept of heterogeneity for thinking about the supply side of strategy. On the supply side, further research might generalize the concept of heterogeneity so as to develop a complementary perspective on strategy to that offered by this thesis. Specifically, strategy can be viewed in terms of two-dimensional heterogeneity at a higher level of abstraction: the creation and exploitation of demand-side *market* heterogeneity and supply-side *resource* heterogeneity.[28] The capability and sustainability of differentiation and segmentation on the market side is inextricably linked with the distinctiveness and imitability of asset structures and production methods on the supply side.[29] The market objective of the strategy of creating differentiated products and insulated customer niches is intertwined with the resource objective of creating—to borrow Rumelt's phraseology—"idiosyncratic capital"[30] that leads to "uncertain imitability"[31] as perceived by competitors.

Competitive replication of distinctive asset base configurations can be inhibited by the characteristics of the investment required. In particular, investment that entails high costs, substantial elapsed time before it reaches fruition, and "irreversibility" that creates substantial risk (in that it exposes firms to irrecoverable losses if the investment does not achieve the desired goal) inhibits imitation. Irrecoverability typically characterizes intangible assets (such as image differentiation or R&D capability) that are difficult to define and that, unlike tangible plants and equipment, have no resale markets.

The distinction between the demand side of strategy and the supply side of strategy (as also used in the introduction of this book) is merely an organizing mechanism useful in clarifying the subject and the scope of this research. Both sides are actually different perspectives on any single strategy. Thus, for instance, IBM's effective exploitation of market heterogeneity in its image as the low-risk vendor entails the creation of resource heterogeneity in that it involves developing a unique and intangible asset that cannot be readily duplicated without high cost, risk and time. Similarly, IBM's marketing strategy of selling solutions, which is dependent on a uniquely configured and deployed field force combining System Engineers with Marketing Representatives, also represents a distinctive asset that requires a substantial and prolonged investment that is not recoverable.

Among niche players the same dual perspective is apparent. For instance, Cray's market strategy of offering only faster and faster computers despite a reduction in their general utility, reliability, and compatibility is also an investment strategy that reflects an unswerving priority on resource allocation to R&D in a highly specialized area. The resultant technological capability is a hard to define asset that cannot be copied without making an irrecoverable commitment of funds to a highly risky, long-term development program. Similarly, the corporate culture cultivated by the unique figure, Seymour Cray, is both an instrument of market strategy and a distinctive asset in that it plays an important role in the company's ongoing capacity for innovativeness necessary to satisfy its customers' evergrowing desire for extreme computing speed. As McGee explains: "It is profitable juxtaposition of idiosyncratic capital against product markets that is the essence of the strategic task."[32]

While subsequent research might be directed vertically to extend the analysis into the supply-side perspective on strategy as suggested above, research can also be directed horizontally by exploring the applicability of the existing demand-side framework in other types of markets. The categories of behavior and their underlying logic as structured in this thesis are applicable to other businesses. For example, there are important parallels between the behavior of IBM and Xerox, particularly concerning complimentary usage of product augmentation and price segmentation and the means for implementing these:

Xerox erected a substantial marketing and product differentiation barrier by its investment in highly trained . . . salesmen, and service and repair engineers [as well as] segregated its customers . . . based upon intensity of usage [and] locked in its customers through a multiplicity of price schedules and discounts [resulting in] discriminatory pricing made possible by . . . higher switching costs.[33]

The framework of market behavior used in this thesis is also applicable to other markets. Work by Fondas found the framework helpful in distilling from behavior the essential strategy of business units in five different industries: conduit fittings, hand tools, window treatments, industrial cables, and oil drilling compressors.[34] McKinsey and Company has also used this framework for diagnosing strategy in a number of businesses: consumer photography, crankshaft forgings, pre-press computer systems, and personal property insurance. It has been particularly useful in highlighting new opportunities for segmented pricing by dominant product augmentors in the insurance business resulting from the unique, but unexploited, characteristic of the insurance industry that identical insurance products can legally be explicitly, differentially priced.

In addition to establishing the potential for wider applicability through the accumulation of fine-grain case-specific research, applicability can be examined through more course-grain approaches applying quantitative methods at a more general level. The first step in applying quantitative analysis to ideas on market strategy is the definition of variables and the compilation of data using those variables. The establishment of a suitable data base will potentially lead to a more precise specification of causal linkages between behavior and outcome variables as well as among behavior variables.

Variables such as product augmentation and product segmentation can be measured and quantified using perceptual data such as is done in conjoint analysis and cluster analysis in marketing as well as employing directly observable data as is typical in hedonic analysis. The resultant data base would thus consist of both perceptual variables and descriptive market statistics. Indeed, the PIMS data base (exceptional if not unique in strategy research) contains perceptual variables such as "product differentiation" as well as objective market-performance variables that may be a useful starting point for quantitative investigations of the general applicability of a more precisely specified extension of the framework presented here.

The framework of two-dimensional market heterogeneity is significant not only for business policy research but also for practice. In particular, the sources of strategic advantage implied by two-dimensional market heterogeneity pose a challenge for conventional product policy practices rooted exclusively in the framework of the "growth-share matrix" (or "product portfolio") and the more fundamental experience curve. In this approach to strategy, the mechanism for gaining competitive advantage is cost reduction through experience and market share growth through price competition.[35]

Regarding the effect of experience and, therefore, market share on costs, innovative product augmentation or product segmentation that results in providing the customer with the same benefits at a lower price can overwhelm the experience effects on cost. As a result, the firm with the largest share is not necessarily the one with the lowest costs. Concerning the effect of cost reduction and consequent price cutting on market share growth, product augmentation that enhances perceived product benefits despite increasing costs to the customer may have greater potential for increasing share than a price reduction on an unaugmented product. Even if this is not the case, the effect on profitability of product augmentation or product segmentation may exceed the combined effect of price reduction on market share and of experience accumulation on firm costs.

The narrowness of such approaches to market strategy lies in the fact that they implicitly ignore the importance of product heterogeneity, customer heterogeneity, information inadequacy, and the potential effectiveness of strategies exploiting these alternative, demand-based sources of market advantage. Moreover, experience accumulation strategies and the more complex product portfolio policies imply a myopic product concept insofar as products are viewed as homogeneous. The result is an incomplete view of the range of strategic options available. For instance, in markets characterized by rapidly changing products where experience effects are overwhelmed by cost reduction due to innovation, product portfolio thinking may overlook the possibility that the more effective approach to gaining and sustaining a profitable market share may lie in the ability to augment the product in ways that, in fact, only add to product cost. Interestingly, in IBM, as in other successful competitors in the mainframe computer industry (which is characterized by rapid process and product change), product portfolio techniques have no role in strategy formation and are not highly regarded.

Furthermore, strategy that exclusively focuses on the accumulation of experience in the production of one product is not tenable as a long-term strategy because it neglects to balance this approach with the need to satisfy heterogeneous customer wants with general and segment-specific product augmentation. More generally, any strategic approach that focuses only on price-competitive tactics and, relatedly, on cost-based advantage is not sustainable over the long term. Such an approach is not tenable as a long-term *sales* strategy because there are limits to the quantity of a homogeneous product consumers are willing to purchase. As Mantell and Sing explain, "regardless of the present volume advantage to be secured through price competition, the present object of expenditure will inevitably give way to a new and different product to satisfy man's desire for variety."[36] Such an approach is also not tenable as a long-term *financial* policy because it provides no means of escaping the profit-reducing tendency of equilibrium competition, or what Levitt calls that "commodity trap" and the "purgatory of price competition."

Long-term firm viability is ultimately related to the ability to temper any focus on the accumulation of experience or, more generally, supply-based cost reduction with the need for demand-based strategy facilitating greater customer satisfaction. As with the implications for welfare and public policy, the significance of two-dimensional market heterogeneity for business policy lies in the importance of the potential incompatibility of efficiency and effectiveness.

The narrow view of strategic options available, implicit in certain strategy techniques, may also be present in scholarly representations of IBM's behavior. This seems to be the case with interpretations of IBM's behavior as primarily a "limit-pricing" strategy.[37] However, price-cutting behavior by IBM does not necessarily represent an attempt to limit entry through lower prices but may simply reflect declining input costs benefiting all firms in its innovative industry.[38] Moreover, spatial economic analysis indicates that "the theory of limit pricing does not apply to the case of differentiated products"[39] because "limit pricing is contrary to profit maximizing behavior in spatial competition.[40] Given the large disincentive for price-based attempts to limit competitive entry created by the threat of self-induced losses, especially in the case of a large rental base, any comprehensive interpretation of IBM's behavior as an entry-limiting strategy must also consider the role of nonprice types of behavior such as interface manipulation in locking out competition. Similarly, any comprehensive interpretation of the behavior of IBM's competitors as entry strategies must not only include their willingness to price under IBM's umbrella as a vehicle to entry but also include the ability to augment products in segment-specific ways as a means to entering the market through an insulated niche.

A strategy directed at exploiting demand-based sources of strategic advantage and thus taking advantage of product heterogeneity, customer heterogeneity, or the potential for reducing customer risk has important implications for organizational structure and management. Efficiency and cost reduction must be balanced against, and integrated with, effectiveness and customer value; in particular, the need to temper efforts toward efficiency with efforts toward effectiveness should be explicitly recognized in a firm's intraorganization operations. This implies that traditionally separate functional areas must manage across conventional boundaries so that the actions of one does not negate the efforts of another. For instance, product design should extend beyond the engineering function so that effectiveness is not lost to efficiency. Similarly, cost control should extend beyond the accounting and finance function. Most importantly, manufacturing must be managed not only for the optimization of efficiency, but also should be consistent with potentially conflicting marketing objectives.[41]

Such organizational implications provide further support for the idea of strategy incompatibilities discussed in Chapter 3 and earlier in this chapter. In particular, potential incompatibility may be present between alternative strategic approaches for organizational reasons separate from the scarcity of firm resources. For example, cost-based strategies and customer value-based strategies may imply different organizational structures, skills, cultures, management styles, and systems.

Attaining cost advantage implies a large role for cost control and thus financial controllership. It suggests rigid planning and control systems, attention to the minimization of overhead, and the elimination of all but the essential cost items. It implies emphasis on the pursuit of scale economies, organizational economies, and learning and experience effects that result from a focus on the production of one standardized product. It suggests a rigid structure with centralized authority, a culture of frugality and attention to detail, and a "mechanistic"[42] system of management.

Such organizational structure, culture, and management behavior may be counterproductive to firms attempting to create superior perceived customer value. Such a firm may wish to subordinate scale economies, learning, and experience effects to the encouragement of experimentation in the creation of new and enhanced products and/or the encouragement of competing projects. An "organic"[43] management style may be more appropriate and an emphasis on cost control might be superceded by the need for built-in redundancies to assure the quality of customer benefits. Such a strategy may imply a less hierarchical structure, with more decentralized authority characterized, for instance, by autonomous profit centers or IBUs (independent business units). Such a strategy may further imply a culture encouraging risk-taking innovators and, therefore, more tolerance toward competing projects and nonessential investments.

SIGNIFICANCE FOR THE STRUCTURALIST PARADIGM

The generic public policy perspective of price competition as an ideal, the tendency to view specifically IBM's behavior in bipolar terms, and the neglect in business strategy thinking of the full dimensionality of market behavior are all rooted in, and grow from, a common feature of the structuralist paradigm as currently conceived. *Given homogeneity,* price competition *is* ideal, behavior is either price competitive or monopolistic, and strategic success lies solely in the ability to gain share through a low-cost position and thus superior price competitiveness. The significance of the idea of two-dimensional market heterogeneity, however, is that it highlights the need for revision of the content of the structural assumptions underpinning the structuralist paradigm of market behavior. What is here challenged is not the generic idea of structuralism per se, but the particular form this structuralism has taken (as summarized in this chapter at the beginning of "Significance for Public Policy"). The feature at issue here is homogeneity, which although no longer presented as a reasonably accurate generalization of observed behavior, does serve as the starting point, the foundation stone, the ideal type upon which theory is built.

In the course of economic and intellectual history, the divergence of actual markets from the assumed condition of homogeneity, in effect, created a dilemma. Rather than remove the assumption of homogeneity, the path intellectual history chose to resolve this problem was to retain the assumption of homogeneity but change it from a "real type" to an "ideal type." Economics thus came to be a study not of actual markets, but of theoretical markets defined in terms of limiting assumptions, and actual behavior was explained as an impure combination of elements of theoretical markets.

The importance of two-dimensional market heterogeneity is that it implies the presence of behavior that is not fully explainable in terms of a combination or alloy of models of behavior based on homogeneity. The endpoints of pure competition and monopoly are inadequate to contain within them a full view of market behavior. Heterogeneous market behavior cannot be fully understood apart from understanding the nature of market heterogeneity, a feature of neither of the supposed theoretical endpoints.

The alternative approach here suggested to reconcile observed behavior with theory is to change the starting point, the ideal type itself, so that the structural assumptions are again close to observed behavior. This is, in a sense, a call to start again and construct new theory based on the heterogeneity of markets rather than amend existing theory based on the theoretical assumption of homogeneity.

This call is inconsistent with the philosophical idea of structuralism. The key objective of structuralism as it originated in its master discipline of linguistics is to reveal the common structure of the reality being studied, as Rousseau first did in the case of language and, in particular, in Indo-European languages.[44] Similarly Chomsky, taking linguistic structuralism a step further, argued that the ultimate contribution of structuralism lies in identifying the few fundamental "deep structures" (rather than the many "surface structures" grammarians had traditionally investigated) from which the whole variety of language is generated.[45] In the field of anthropology, Radcliffe-Brown added structuralism to the concept of function as found in Durkheim and Malinowski to create the notion of the structure of a functioning system.[46] In this view, structure is not merely form, but the "significative" content of the system. Structure is the reason systems function the way they do. It is only this actual, significative, deep structure that allows us to develop valid generalizations about observable phenomenon, or as anthropologist Levi-Strauss pointed out, allows us to "draw upon an inventory of mental patterns, to reduce apparently arbitrary data to some kind of order."[47]

Structuralism as applied to firm behavior in markets should follow the lead of its intellectual forefathers. Indeed the philosophical concept of structuralism is readily applied to firm behavior in markets. The "system" is the market system. The function, *in the conventional view,* is efficiency, which logically follows from the structural assumption of homogeneity. But is this structural feature representative of the "deep structure" or "significative content" of the market system? Or is it merely a convenient fiction adopted to avoid muddling the elegance of existing theory?

This book argues that this structural assumption should be changed and made more complex to include heterogeneity and indeed emphasize heterogeneity as a prevailing structural feature. Consistent with the philosophical idea of structuralism, modification of the structural base implies modification of the function of the system. From the structural assumption of two-dimensional heterogeneity flow a less simple view of the function of markets to include matching products and customers, through the superordinate mechanism of market information, toward the potentially conflicting welfare objectives of both efficiency and effectiveness.

The work goes beyond a call to recognize that "heterogeneity" leads to the importance of nonprice behavior in markets. Because it proposes the more elaborate construct of "two-dimensional market heterogeneity," it identifies not only the importance of nonprice behavior but also the fundamental anatomy of that behavior. This anatomy of two-dimensional heterogeneous market behavior provides a foundation for further theoretical and applied research. It is a structural skeleton upon which theory can be fleshed out, and it is a heuristic for dissecting actual, observed behavior, thereby revealing its fundamental strategic logic. Further research may confirm the applicability of this framework of two-dimensional heterogeneous market behavior to different specific markets. This book shows that it is useful in describing, understanding, and diagnosing strategy in the market for computer mainframe products.

NOTES

1. The very phrase "consumer's surplus" betrays the negative effect on welfare associated with price discrimination. The word "consumer's surplus" suggest that "the surplus rightfully belongs to the consumer for having obeyed the law of supply and demand" (L. H. Mantell, F. P. Sing, *Economics for Business Decisions,* [New York, McGraw Hill, 1972], p. 26). Indeed, the most prominent early work of price discrimination was cast within the broader context of "the economics of welfare" and specifically analyzed price discrimination in terms of the three degrees to which it unfairly extracted consumer's surplus; see Pigou, *The Economics of Welfare,* p. 277.

2. For a mathematical discourse on optimal product variety, see Lancaster, *Variety, Equity, and*

Efficiency; also see M. Spence, "Product Differentiation and Welfare," *American Economic Review,* 66 (May, 1976), pp. 407–414.

3. F. M. Scherer, "The Welfare Economics of Product Variety, *Journal of Industrial Economics* (December, 1979), pp. 113–134; also see Scherer's p. 394.

4. For an analysis of the relationship between optimal product variety and price discrimination see M. Spence, "Product Selection, Fixed Costs, and Monopolistic Competition," *Review of Economic Studies,* 43 (June, 1976), pp. 218–220.

5. G. Norman, "Spatial Competition and Spatial Price Discrimination," *Review of Economic Studies,* 48, No. 1 (January, 1981), pp. 97–111.

6. Empirical work by Greenhut supports the proposition that greater transport cost absorption occurs at those points in the market where competition is more intense. J. Greenhut, M. L. Greenhut, S. Li, "Spatial Pricing Patterns in the United States," *Quarterly Journal of Economics,* 94, No. 2 (March, 1980), pp. 329–350.

7. Hobbs, "Mill Pricing Versus Spatial Price Discrimination," p. 189.

8. N. Dorward, "Recent Developments in the Analysis of Spatial Competition and Their Implications for Industrial Economics," *The Journal of Industrial Economics,* 31, Nos. 1–2 (September–December, 1982), p. 142.

9. P. Asch, *Industrial Organization and Anti-Trust Policy* (New York, Wiley, 1983), p. 98.

10. R. G. Lipsey, K. Lancaster, "The General Theory of Second Best," *Review of Economic Studies,* 24 (1956), pp. 11–32; also see P. Bohm, "On the Theory of Second Best," *Review of Economic Studies,* b34 (July, 1967), pp. 301–314.

11. W. S. Block, "Preface," in D. Armstrong, *Competition Versus Monopoly: Combines Policy in Perspective* (Vancouver, B.C., The Fraser Institute, 1982), pp. xix–xxiv.

12. Department of Justice, "Merger Guidelines" (31 May 1986), p. 2.

13. C. J. Colwell, "One of the Congeries of Anti-Competitive Practices," *The Southern Economic Journal,* 33, No. 4 (April, 1967), pp. 546–547.

14. L. W. Stern, and J. R. Grabner, Jr., *Competition in the Marketplace* (Glenview, IL, Scott Foresman, 1970), p. 6.

15. Block, "Preface" in *Competition Versus Monopoly,* p. xvii.

16. D. R. Capozza, R. Van Order, "A Generalized Model of Spatial Competition," *American Economic Review,* 68 (1978), pp. 896–908, as summarized by Greenhut, Nopman, Hung, *The Economics of Imperfect Competition,* p. 4.

17. Greenhut, Nopman, Hung, *The Economics of Imperfect Competition, A Spatial Approach,* p. 2.

18. Dorward, "Recent Developments in the Analysis of Spatial Competition," p. 149.

19. Brock, *The U.S. Computer Industry.*

20. DeLamarter, *Big Blue.*

21. F. M. Fisher, J. J. McGowan, J. W. Greenwood, *Folded, Spindled, and Mutilated.*

22. F. M. Fisher, J. W. McKie, R. B. Manke, *IBM and the U.S. Data Processing Industry.*

23. Armstrong, *Competition Versus Monopoly,* p. 49.

24. Brock, *The U.S. Computer Industry,* pp. 183–223.

25. M. E. Porter, "The Technological Dimension of Competitive Strategy," in R. Rosenbloom, *Research on Technological Innovation, Management and Policy* (Greenwich, CT, J.A.I. Press, 1987).

26. H. Mintzberg, "Patterns in Strategy Formation," *Management Science,* 24, No. 9 (May, 1978), pp. 934–948. See also H. Mintzberg, "Strategy Making in Three Modes," *California Management Review* (Winter, 1973).

27. The term "anomaly" is used here specifically in the Kuhnian sense as an observable phenomenon inconsistent with the prevailing paradigm, thereby revealing the limitations of that paradigm; see T. S. Kuhn, "The Structure of Scientific Revolution," 2d ed. (Chicago, University of Chicago Press, 1970), pp. 52–65.

28. From the perspective of business strategy research this heterogeneity is not simply an exogenous property of markets or industries but is created, manipulated, and exploited through "discretionary," "deliberate," "un-innocent," "strategic" activity.

29. Obstacles to competitive imitation can be industry specific ("entry barriers") group specific ("mobility barriers"), or firm specific ("isolating mechanisms"). For the seminal work on "entry barriers," see J. S. Bain, *Barriers to New Competition.* For "mobility barriers," see R. E. Caves, M. E.

Porter, "From Entry Barriers to Mobility Barriers: Conjectural Decisions and Contrived Deterence to New Competition," *Quarterly Journal of Economics,* 91 (1977), pp. 241–262; also see J. McGee, H. Thomas, "Strategic Groups: Theory, Research, and Taxonomy," *Strategic Management Journal,* 7 (1986), pp. 141–160. For "isolating mechanisms," see R. P. Rumelt, "Towards a Strategic Theory of the Firm," in *Competitive Strategic Management,* edited by R. B. Lamb (Englewood Cliffs, NJ, Prentice-hall, 1984), pp. 556–571.

30. Rumelt, "Towards a Strategic Theory of the Firm," in *Competitive Strategic Management,* pp. 566–571.

31. S. A. Lippman, R. P. Rumelt, "Efficiency Differentials Under Competition: A Stochastic Approach to Industrial Organization," University of California at Los Angeles (1981).

32. J. McGee, "Barriers to Growth for Small Innovating Companies in the U.K. and the Role of Competitive Strategy: Effects of Market Structure," Templeton College Management Research Papers, Oxford University (June, 1987), p. 14.

33. A. Ghazanfar, J. McGee, H. Thomas, "The Impact of Technological Change on Industry Structure and Corporate Strategy: The Case of the Reprographics Industry in the United Kingdom," Templeton College Management Research Papers, Oxford University (July, 1986), pp. 11, 12.

34. Fondas, "Managerial Agendas and Strategic Alignment."

35. For an explication of this conventional approach to strategy, see two articles by H. Hedley, Director of Braxton's London Office: "A Fundamental Approach to Strategy Development," *Long Range Planning* (December, 1976), pp. 2–11, which focuses on the experience curve framework, and "Strategy and the 'Business Portfolio,' " *Long Range Planning,* 10 (February, 1977), pp. 9–15. For a seminal criticism of specifically the "product portfolio," see G. S. Day, "Diagnosing the Product Portfolio," *Journal of Marketing,* 41, No. 2 (April, 1977), pp. 29–38.

36. Mantell, Sing, *Economics for Business Decisions,* p. 26.

37. See for instance J. Sengupta, E. Leonard, J. P. Vanyo, "A Limit Pricing Model for U.S. Computer Industry: An Application," *Applied Economics* (June, 1983), pp. 297–308. For an earlier application of the limit pricing framework to the computer industry, see Geruson, "Elements of Strategy in IBM's Price and Product Behaviour," pp. 17–22, 28–34.

38. D. A. Hay, "Sequential Entry and Entry-Deterring Strategies in Spatial Competition," *Oxford Economic Papers,* 28, No. 2 (July, 1976), p. 253.

39. Dorward, "Recent Developments in the Analysis of Spatial Competition," p. 146.

40. Geruson, "Elements of Strategy in IBM's Price and Product behaviour," p. 21.

41. See, for instance, B. P. Shapiro, "Can Marketing and Manufacturing Coexist?" *Harvard Business Review,* 56 (September–October, 1977), pp. 104–114.

42. T. Burns, G. M. Stalker, *The Management of Innovation* (London, Travistock Publications, 1981), pp. 120–122.

43. Ibid.

44. J. J. Rousseau, *On the Origin of Language* (Chicago, University of Chicago Press, 1986).

45. N. Chomsky, *Language and Mind* (New York, Harcourt Brace Jovanovich, 1972).

46. A. R. Radcliffe-Brown, *Structure and Function in Primitive Society: Essays and Addresses* (New York, Free Press, 1965).

47. C. Levi-Strauss, *Mythologiques,* 2 (New York, Harper and Row, 1973).

Appendix A

Research Strategy

This study is an "archival analysis."[1] Among the most important components of a research design[2] in archival analysis, as with other types of research, are: (1) the subject of analysis, (2) the unit of analysis, (3) the types of evidence, and (4) the sources of evidence. The subject of analysis consists of the boundaries of the conceptual subject of the study, how this conceptual subject is "operationalized" for empirical research, and the boundaries of the empirical subject. These factors, combined with the unit of analysis, form the basis for articulating the research question that in this study is addressed with multiple types and sources of evidence.

THE CONCEPTUAL SUBJECT OF ANALYSIS

The conceptual subject of this study is "market strategy," which comprises a subfield contained within the field of business strategy. The concept of "market strategy" as defined in this study derives from the broader idea of "business strategy," well established in business literature. The common essence of various manifestations of "strategic business behavior" and the essential meaning of business strategy lies in the posture of a company with respect to its environment and more specifically involves "developing and maintaining a viable relationship between the organization and its environment."[3] Modifying the idea of strategy as the relation between the firm and its environment by the concept of market as a "place" where sellers sell products to customers results in the more specific concept of market strategy as the relation of a firm and its products to its *customer* environment. Similarly, given the idea of effectiveness as the ability to satisfy a firm's goals with its output, "*market* effectiveness" is the ability of a firm to achieve the goal of customer satisfaction using its product offerings.[4]

It is important to distinguish the research subject of *market* strategy from the related but distinct subfield of "*competitive* strategy." These two are distinguished in their environmental foci, their immediate objective function, and as developed in this study, their underlying logic. Competitive strategy focuses on the competitive environment ("rivalry") and attempts to improve business performance by improving the posture of a firm in "the context of identifiable adversaries."[5] Market strategy, on the other hand, focuses on the customer environment and attempts to improve business performance by better positioning the firm and its product offerings with respect to its customers. The notion of firm "conduct" as used in the competitive strategy framework refers to conduct with respect to *competitors*. Conduct, or what is in this study called "market behavior," refers to the activity of firms in the context of *customers*. Even more specifically, as Christian and Hofer argue, the term "differentiation" as used in the competitive-strategy framework, refers ultimately to "differentiation with respect to competitors."[6] Differentiation as used in this study refers to differentiation of products as perceived by customers.

In addition to its environmental foci, and immediate objective function, market strategy differs from competitive strategy also in its underlying logic and its basis for identifying sources of strategic advantage. The underlying logic and basis for identifying sources of "competitive advantage" in the competitive strategy area rest on the methodological approach and analytic logic of industrial organization (IO) economics originating with Mason[7] and Bain,[8] as further developed by Scherer[9] and Caves,[10] and as adapted to the objective function of business strategy by Porter.[11] Market strategy, as here conceived, is rooted in a more narrow and fundamental economic logic.

The competitive strategy framework adopts structure-conduct-performance logic[12] to the business objective of competitive advantage (improving position within the competitive environment) and specifically looks to the exploitation of industry structure for sources of competitive advantage. This book proposes that a more basic economic logic can be adopted to the different business objective of market effectiveness. In particular, the logic that flows from demand analysis in economics is useful for developing a framework of customer-focused market behavior because the demand curve model is rooted in the concept of "utility" or, in other words, the satisfaction customers derive from the purchase of products. More specifically, market strategy (the conceptual subject of this study) as here defined, focuses on demand-based sources of improved business performance. It is therefore distinct from supply-based sources of strategic advantage, such as the selection of production methods, which can lead to experience curve effects or economies of scope.[13]

Operationalizing the Conceptual Subject for Empirical Research

This concept of market strategy must be further defined so as to be operationally useful in empirical research. Here again, ideas on the broader subject of business strategy are helpful. The conventional definition of business strategy as what "the company intends . . . its objectives, purposes or goals"[14] is methodologically inadequate for empirical research. To this notion of strategy must be added the idea of strategy as an "emergent pattern" in behavior.[15] As Mintzberg explains:

By restricting strategy to explicit, a priori guidelines, it forces the researcher to study strategy . . . as a perceptual phenomenon [with the result that] the literature on strategy . . . is in a large part theoretical and not empirical. . . . The position taken here is that this definition is incomplete for the organization and non-operational for the researcher.[16]

By including in the definition of strategy the observable patterns in actual behavior (product changes, pricing programs, and marketing techniques used by the seller, and perceptions and purchasing behavior of the customers), strategy is identifiable irrespective of empirical evidence of intent. Indeed, to the extent that "logical incrementalism" characterizes strategy formation, such patterns may "give the appearance of consciously formulated strategy but may be the natural result of compromise among coalitions backing contrary policy proposals or skillful improvisatory adaptation to external forces."[17] The primary form of observing *market* strategy in this study, therefore, is by discerning the patterns in actual behavior that relate to products and customers: product actions, pricing programs, and marketing techniques used by the seller and the purchasing behavior of customers.

For purposes of empirical strategy research this is generally the only methodologically practicable definition of strategy. However, throughout the applied part of this thesis, the use of actual behavior as evidence is supplemented with quotations from court testimony by IBM executives and internal "IBM-confidential" memos, reports, and meeting minutes publicly disclosed as a result of various private and public lawsuits brought against IBM, especially the most recent suit brought by the U.S. Justice Department. This additional data provides an unusual opportunity for observing intended strategy.

The Empirical Subject of Analysis

The *empirical* subject of this study is primarily IBM's intended and emergent behavior in the market for mainframe computer products and secondarily the behavior of a few selected niche competitors in the same market. The products in this market are the central processing units (CPUs) and also the peripheral devices that attach to CPUs, such as tape drives, disk drives, and terminals. The sources of data dictated the geographical boundaries of the market studied as those of IBM's domestic market: the United States.

Unit of Analysis

The unit of analysis of this study is the business unit, although the analysis at times reaches beneath this level to the policies and structure of the marketing function within the firm. It is important to distinguish this level of analysis from the related fields of corporate strategy and marketing, the analysis levels of which (the corporate entity and the business function) bound that of market strategy (the business unit). The objective function of corporate strategy—to improve the performance of a corporate entity consisting of one or more businesses and, more specifically, to improve value to shareholders or other stakeholders[18]—subsumes the more narrow goal of market effectiveness. Hence, corporate strategy includes a wide range of behavior that exceeds the parameters of market strategy, such as restructuring to improve corporate value, diversification (related or unrelated/vertical or horizontal), and the management of a portfolio of business units.

The field of marketing on the other hand, although closely related to market strategy, is distinguished from it because it focuses on phenomena at the level of the business function. The study of the function of marketing comprises a broad, eclectic, interdisciplinary field, the soft boundaries of which extend to a wide range of specific functional-level activity (such as sales force incentives, distribution programs, and advertising techniques) that exceeds the purview of this study.

Types of Evidence

This study uses "multiple sources of evidence [to facilitate] the development of converging lines of inquiry, a process of triangulation."[19] The evidence includes various types of numerical and nonnumerical information; therefore, the research strategy of this study should not be confused with the use of exclusively "qualitative research."[20] The types of evidence include survey data, financial data and descriptive data in different forms. More specifically, the evidence includes statistical survey data on customer purchasing behavior; financial data as to prices, costs, and profit margins of product offerings; and numerical and nonnumerical data describing product actions, product characteristics, product performance, customer behavior, customer characteristics, marketing techniques, as well as structural and behavioral features of the marketing function.

The evidence also consists of both historical data and the description of contemporary phenomena. The use of such data, however, should not be misinterpreted to imply this study is either a business history or a simple case study; it is, rather, an archival analysis drawing on types of evidence that are also used in these alternative, related types of research.

Sources of Evidence

The major sources of data are archival and include economic, business, historical, and legal descriptions and analyses of competitors and customers in the market for mainframe products found in books, internal company reports, independent private research reports, court documents, scholarly periodicals, trade journals, business press articles, and case research at business schools. The most important sources of archival data used are a recently compiled, detailed statistical data base of customer purchasing behavior and the *U.S. v. IBM* court case disclosing previously confidential, internal firm information revealing both intentions and patterns in actual market behavior. In addition to these archival sources, data were collected through field research in the form of unstructured interviews with IBM employees in various job positions and at different job levels, as well as interviews with several industry experts in consulting firms, investment firms, and graduate business schools.[21] Field research was further facilitated through "insider" observation as a result of the author's employment in the strategy planning and pricing functions at IBM.

Research Questions

It is frequently the case in studies of this sort to formulate and articulate one or more research questions that attempt to capture the essence of the research. This study addresses the question of how aspects of IBM's behavior, and specifically the actions relating to its customer environment, improve its market posture. This general question can be divided into two parts. The study seeks to discern patterns in IBM's market behavior and, therefore, reveal elements of IBM's market strategy. Further, it seeks to discern the underlying logic of this behavior and thus reveal how IBM's behavior contributes to an advantageous market posture; in other words, it seeks to identify sources of IBM's market effectiveness. The same questions are addressed for a few other selected competitors in the mainframe market.

The general findings of this research are that distinct patterns are present in IBM's behavior that reveal elements of its market strategy. These can be accounted for in terms of different areas of economic logic and marketing concepts that help explain how such behavior contributes to superior business performance and an advantageous market posture. These previously distinct, generic theoretical ideas can be organized into a new, single framework that addresses the further question of what the behavior of IBM (and others) tells us about sources of market effectiveness.

Research Process

These research questions and findings, however, tend to oversimplify the subject of research and hide the process of research leading to the development of the research subject. Initially this study addressed the question of IBM's behavior and its underlying logic. This led to identification of two sources of IBM's effectiveness (product augmentation and price segmentation) and the development of a conceptual heuristic for organizing them. This conceptual device in turn revealed the possibility of two other generic sources of effectiveness (product segmentation and information inadequacy). The applicability of these predicted categories was confirmed by further observation of IBM's behavior and that of a few other appropriate mainframe competitors. In short, the conceptual heuristic developed from one set of observations drove further observation. The book thus, not only explores behavior to identify strategy and its underlying logic but also shows the applicability of a heuristic derived from this logic.

NOTES

1. R. K. Yin, *Case Study Research: Design and Methods* (Beverly Hills, CA, Sage, 1986), pp. 13–26.

2. Ibid., p. 29.

3. P. Kotler, *Marketing Management: Analysis, Planning and Control,* 4th ed. (London, Prentice Hall, 1980), p. 65.

4. Bonoma, *The Marketing Edge,* pp. 191–199.

5. R. Howitz, "Policy, Strategy, and Planning—A Reformulation" (London, London Regional Management Center Occasional Paper, 1976), p. 2.

6. J. J. Chrisman, W. R. Boulton, C. W. Hofer, "A Conceptual Note on Generic Strategies and Strategy Classification Systems," *Southern Management Association Proceedings,* edited by D. F. Ray (Southern Management Association, 1986), p. 223.

7. E. S. Mason, "Price and Production Policies of Large-Scale Enterprise," *American Economic Review,* Supplement 29 (March, 1939), pp. 61–74, and "The Current State of the Monopoly Problem in the United States," *Harvard Law Review,* 62 (June, 1949), pp. 1265–85, by the same author.

8. J. S. Bain, *Barriers to New Competition* (Cambridge, MA, Harvard University Press, 1956); see also *Industrial Organization* (New York, Wiley, 1959) by the same author.

9. Scherer, *Industrial Market Structure.*

10. R. E. Caves, *American Industry: Structure, Conduct, Performance,* 4th ed. (Englewood Cliffs, NJ, Prentice Hall, 1977).

11. M. E. Porter, "How Competitive Forces Shape Strategy," *Harvard Business Review,* (March–April, 1979), pp. 137–145, *Competitive Strategy, Competitive Advantage,* by the same author.

12. M. E. Porter, "The Contribution of Industrial Organization to Strategic Management: A Promise to be Realized," *Academy of Management Review* (October, 1981) and "Industrial Organization and the Evolution of Concepts for Strategic Planning: The New Learning," *The Integration of Corporate Planning Models and Economics,* edited by T. H. Naylor (Amsterdam, North Holland Publishing, 1982).

13. P. Ghemewat, "Sustainable Advantage," *Harvard Business Review,* 64, No. 5 (September–October, 1986).

14. C. R. Christensen, K. R. Andrews, J. L. Bower, R. G. Hamermesh, M. E. Porter, *Business Policy,* 5th ed. (Homewood, IL, Irwin, 1982), p. 93.

15. H. Mintzberg, "An Emerging Strategy of Direct Research," *Administrative Science Quarterly,* 24 (December, 1979), p. 582.

16. Mintzberg, "Patterns in Strategy Formation."

17. J. B. Quinn, "Strategic Change: Logical Incrementalism," *Sloan Management Review* (Fall, 1978), p. 97.

18. H. I. Ansoff, *Corporate Strategy: An Analytical Approach to Business Policy for Growth and Expansion* (New York, McGraw Hill, 1965), pp. 32–34.

19. Yin, *Case Study Research,* p. 91. See also T. V. Bonoma, "Case Research in Marketing: Opportunities, Problems and a Process Model," Revision 3.4 (unpublished paper, Harvard Business School, November, 1984).

20. See J. Van Maanen, J. M. Dobbs, R. R. Faulkner, *Varieties of Qualitative Research* (Beverly Hills, CA, Sage, 1982), p. 16.

21. The interviews are conducted at McKinsey & Company; Booz, Allen and Hamilton; The Boston Consulting Group; Bain and Company; The Hay Group; Monitor; Braxton Associates; Braxton Development Corporation; the Sloan School of Management of the Massachusetts Institute of Technology; and the Harvard Business School.

Appendix B

Spatial Models of Competition and the Tendency toward Differentiation

Hotelling's widely cited conclusion, in literal terms, was that successive locational moves by two firms will lead to a single, stable, long-run equilibrium, with each firm back to back at the center of the market.[1] Such "pairing" or "clustering" of firms in the literal sense of the model translates into a tendency for competition to result in a reduction of product differentiation to some minimum level. Some writers initially conjectured that the tendency for clustering (or "Principle of Minimum Differentiation," as it has become known) is a pervasive equilibrium property of more general spatial models. However, subsequent research shows that the result is highly sensitive to changes in the assumptions that often lead to a "dispersion" of firms, in other words, wide product variety.

Subsequent work can be categorized by the features it introduces. Among the most important features introduced are:

- More than two firms
- Elastic demand
- Seller relocation costs
- Agglomeration economies
- Circular market configurations
- Two-dimensional market space

The results of research in each of these areas vary both within and among categories. In particular, no clear consensus has emerged about the tendency for, and level of, clustering (or dispersion) or the existence and uniqueness of equilibria.

For instance, Lerner and Singer found that the introduction of four or more additional firms into Hotelling's model (six or more in total) results in some clustering but no unique equilibria.[2] Smithies found that where consumer demand is price elastic (a relaxation of Hotelling's assumption), the gains from encroaching on a rival duopolist's location are offset by the loss of sales due to higher transport costs, thereby eliminating the necessity of the pairing result.[3] Hay found that the introduction of seller relocation costs, to the point where leapfrogging is prohibitively costly, encourages the firm to locate so as to minimize the risk of enforced relocation. The process of locating to deter entry results in a scattering of firms throughout the market.[4] The introduction of agglomeration economies can create an "incentive for clustering"[5] but also can lead to "symmetric dispersed equilibria" whereby firms have distinct locations.[6] The specific economies of agglomeration that result from "a spatial concentration of demand," which in turn is motivated by the "reduction in search costs" to customers, can make clustering more profitable to the seller than his securing "maximum monopolistic advantage by increasing search costs as much as possible and therefore choosing locations distant form each other."[7]

Substituting the circumference of a circle for a bounded line to represent the market provides similarly variable results. The absence of endpoints on a circle removes the disparity between firms at the end of the line, with only one neighbor, and the other firms, with two neighbors. More generally, the circle destroys neighboring relations between some points, thereby changing the assumed, implicit structure of preferences.[8] Novshek finds that by using a circular market, equilibrium exists where it may not exist otherwise.[9] Anderson, on the other hand, finds that a circular model leads to equilibrium only under a very restrictive set of circumstances.[10]

Fully developed models using two-dimensional space suggest a tendency toward dispersion. The Loschian model originating in location theory leads to a positive but bounded level of firm dispersion (or product differentiation). Specifically, the corollary of the model's equilibrium, in physical space, is the arrangement of firms in an hexagonal pattern covering the market.[11] Empirical analyses using a two-dimensional market find an incentive for dispersion because clustering tends to leave profitable entry opportunities in the wide spaces with no firms.[12]

In a comprehensive study examining a range of market configurations, Eaton and Lipsey found that the tendency for differentiation varies depending on a variety of factors and that no conclusive generalization can be reached. However, in contrast to other work, they tentatively suggested that a wide range of equilibrium and disequilibrium situations involve clusters (in two-dimensional space) and pairs (in one-dimensional space) scattered over the market.[13]

NOTES

1. Actually Hotelling's conclusion has been found to be not only not proved but also incorrect. See C. D. Aspremont, J. Gabszewicz, and J. F. Thisse, "On Hotelling's 'Stability in Competition,'" *Econometrica*, 47 (1979), pp. 1145–50; D. Graitson "Spatial Competition a la Hotelling: A Selective Survey," *Journal of Industrial Economics*, 32, Nos. 1–2 (September–December, 1982), p. 11.

2. A. Lerner, H. Singer, "Some Notes on Duopoly and Spatial Competition," *Journal of Political Economy*, 45, No. 2 (April, 1937), pp. 146–186.

3. A. Smithies, "Optimum Location In Spatial Competition," *Journal of Political Economy*, 49, No. 3 (June, 1941), pp. 423–439.

4. Hay, "Sequential Entry and Entry-Deterring Strategies in Spatial Competition," pp. 240–257.

5. G. M. P. Swann, "Product Competition in Microprocessors," *Journal of Industrial Economics*, 24, No. 1 (September, 1985), p. 36.

6. A. de Palma, V. Ginsburgh, J. Thisse, "On the Existence of Location Equilibria in the 3-Firm Hotelling Problem," *Journal of Industrial Economics*, 36, No. 2 (December, 1987), p. 245.

7. K. Stahl, "Differentiated Products, Consumer Search and Locational Oligopoly, *Journal of Industrial Economics*, 23, Nos. 1–2 (September–December, 1982), p. 97.

8. I. Horstmann, A. Slivinski, "The Foundations of Location Models of Product Differentiation" (University of Western Ontario Economics Department, 1982).

9. W. Novshek, "Equilibrium In Simple Spatial (Or Differentiated Product) Models," *Journal of Economic Theory*, 22 (1980), pp. 313–326.

10. S. P. Anderson, "Equilibrium Existence in the Circle Model of Product Differentiation" (*London Papers in Regional Science*), 16 (1986), p. 19.

11. A. Losch, *The Economics of Location* (New Haven, Yale University Press, 1954).

12. F. M. Scherer, "The Welfare Economics of Product Variety: An Application to the Ready-to-Eat Cereals Industry," *Journal of Industrial Economics*, 28, No. 2 (December, 1979), pp. 113–134; see also R. Schmalensee, "Entry Deterence in the Ready-to-Eat Breakfast Cereal Industry," *Bell Journal of Economics*, 9, No. 2 (Autumn, 1978), pp. 305–327.

13. B. C. Eaton, R. G. Lipsey, "The Principle of Minimum Differentiation Reconsidered: Some New Developments In the Theory of Spatial Competition," *Review of Economic Studies*, 42, No. 129 (1975), pp. 27–50.

Appendix C

Net Income before Tax for U.S. Data Process Operations

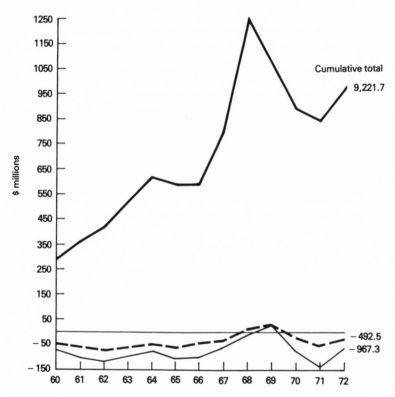

*Burroughs, Sperry Univac, NCR, Control Data Corp., and Honeywell.
**The B.U.N.C.H. plus RCA, General Electric, Digital Equipment Corp., and SDS/Xerox.

——— IBM
— — — The B.U.N.C.H.*
——— Total selected competitors**

Appendix C. Net income before Tax for U.S. data process operations (for IBM and selected competitors). *Source:* Extracted from data compiled by R. T. DeLamarter from a variety of accounting records. See *Big Blue: IBM's Use and Abuse of Power,* pp. 351, 352.

Appendix D

IBM 360 CPU Submodels Included in Each "Submodel Type" Category in Figure 5.4

	Bottom of the Line	Lower End	Middle	Higher End	Top of the Line
	20-1 B		20-1 C		20-1 D
	20-2 B		20-2 C		20-2 D
	30 C	30 D		30 E	30 F
	40 C	40 E	40 F	40 G	40 H
	50 C	50 G		50 H	50 I
	65 G	65 H		65 I	65 J
Total revenue ($ million)	$81.7	$518.9	$449.5	$1819.6	$1158.4
Total profits ($ million)	$14.5	$64.5	$93.3	$559.7	$430.2
Margin	17.7%	12.4%	20.8%	30.8%	37.1%

Source: IBM Blue Letters and *U.S. v. IBM*, PX 1962.

Selected Bibliography

Adams, W. J., Yellen, J. L., "Commodity Bundling and the Burden of Monopoly," *Quarterly Journal of Economics* (August, 1976), pp. 475–498.

Aguilas, F. J., Brainard, C., "Cray Research Inc.," available from Harvard Business School Case Services, No. 385-001 (Boston, June, 1986).

Akerlof, G., "The Market for Lemons: Quality Uncertainty and the Market Mechanism" *Quarterly Journal of Economics*, 84, No. 3 (August, 1970), pp. 488–500.

Alderson, W., *Marketing Behavior and Executive Action: A Functionalist Approach to Marketing Theory* (Homewood, IL, Irwin, 1957).

Alderson, W., *Dynamic Marketing Behavior: A Functionalist Theory of Marketing* (Homewood, IL, Irwin, 1965).

Ansoff, H. I., *Corporate Strategy: An Analytical Approach to Business Policy for Growth and Expansion* (New York, McGraw Hill, 1965).

Anthony, R. N., *Management Control in Non-Profit Organizations*, 3d ed. (Homewood, IL, Irwin, 1981).

Armstrong, D., *Competition Versus Monopoly: Combines Policy in Perspective* (Vancouver, BC, The Fraser Institute, 1982).

Arndt, J., "On Making Marketing Science More Scientific: Role of Orientations, Paradigms, Metaphors and Puzzle Solving," *Journal of Marketing*, 49 (Summer, 1985), pp. 11–23.

Assail, H., "Demand Criteria for Normative Market Segmentation Theory: A Retrospective View," in D. M. Gardner, F. W. Winer, eds., *Proceedings of the 11th Paul D. Converse Symposium* (Chicago, American Marketing Association, 1982), pp. 8–18.

Atwater, T. V., "'Lost' or Neglected Components of a General Equilibrium Theory of Marketing," in P. C. Ferrell, S. W. Brown, C. W. Lamb, Jr., eds., *Conceptual and Theoretical Developments in Marketing* (Chicago, American Marketing Association, 1979), pp. 184–196.

Block, W., "Preface," in D. Armstrong *Competition Versus Monopoly: Combines Policy in Perspective* (Vancouver, BC, The Fraser Institute, 1982), pp. xv–xxix.

Bonoma, T. V., *The Marketing Edge: Making Strategies Work* (New York, Free Press, 1985).

Bonoma, T. V., Shapiro, B. P., *Segmenting the Industrial Market* Lexington, MA, Lexington Books, 1984).

Brock, G. W., *The U.S. Computer Industry: A Study in Market Power* (Cambridge, MA, Ballinger, 1975).

Brock, G. W., "A Study of Prices and Market Shares in the Computer Mainframe Industry: Comment" *Journal of Business*, 52, No. 1 (1979), pp. 119–121.

Burns, T., Stalker, G. M., *The Management of Innovation* (London, Tavistock Publications, 1961).

Cassidy, R., Jr., "Techniques and Purposes of Price Discrimination," *Journal of Marketing* (October, 1946), pp. 135–140.

Caves, R. E., Williamson, P. J., "What is Product Differentiation, *Really?*" *Journal of Industrial Economics*, 34, No. 2 (December, 1985), pp. 113–131.

Chamberlain, E. H., *The Theory of Monopolistic Competition: A Reorientation of the Theory of Value*, 7th ed. (Cambridge, MA, Harvard University Press, 1965).

Chrisman, J. J., Boulton, W. R., Hofer, C. W., "A Conceptual Note on Generic Strategies and Strategy Classification Systems," in D. F. Ray ed., *Southern Management Association Proceedings* (Southern Management Association, 1986), pp. 222–224.

Christensen, C., Andrews, K. R., Bower, L., Hamermesh, R. G., Porter, M. E., *Business Policy*, 5th ed. (Homewood, IL, Irwin, 1982).

Corey, R. E., *Industrial Marketing* (Englewood Cliffs, NJ, Prentice-Hall, 1983).

Cox, D. F., *Risk Taking and Information Handling in Consumer Behavior* (Boston, Harvard University, Graduate School of Business Administration, Division of Research, 1967).

Day, G. S., "Diagnosing the Product Portfolio," *Journal of Marketing*, 41, No. 2 (April, 1977), pp. 29–38.

DeBruicker, F. S., Summe G. L., "Make Sure Your Customers Keep Coming Back," *Harvard Business Review*, 85, No. 1 (January–February, 1985), pp. 92–98.

DeFillipi, R. J., "Causal Ambiguity, Barriers to Imitation, and Sustainable Competitive Advantages," *Academy of Management Review*, 15, No. 1 (January, 1990).

DeLamarter, R. T., *Big Blue: IBM's Use and Abuse of Power* (New York, Dodd Mead, 1986).

Demsetz, H., *Economic Legal and Political Dimensions of Competition: Professor Dr. F. DeVries Lectures in Economic Theory* (New York, North-Holland, 1982).

Dickson, P. R., Ginter, J. L., "Market Segmentation, Product Differentiation, and Marketing Strategy," *Journal of Marketing*, 51, No. 2 (April, 1987), pp. 1–10.

Douglas, E., *Economics of Marketing* (New York, Harper and Row, 1975).

Enis, B. M., Cox, K. K., *Marketing Classics: A Selection of Influential Articles*, 3d edition (Boston, Allyn and Bacon, 1977).

Ferrell, S. W., Brown, S. W., Lamb, C. W. Jr., *Conceptual and Theoretical Developments in Marketing* (Chicago, American Marketing Association, 1979).

Fisher, F. M., McGowan, J. J., Greenwood, J. E., *Folded, Spindled and Mutilated: Economic Analysis and U.S. vs. IBM* (Cambridge, MA, The MIT Press, 1983).

Fisher, F. M., McKie, J. W., Manke, R. B., *IBM and the U.S. Data Processing Industry: An Economic History* (New York, Praeger, 1983).

Fishman, K. D., *The Computer Establishment* (New York, Harper Row, 1981).

Fombrum, C., Shanley, M., "What's In a Name? Reputation Building and Corporate Strategy. *Academy of Management Journal;*" 33, No. 2 (June, 1990).

Fondas, N. J., "Managerial Agendas and Strategic Alignment: A Field Study of General Management Behavior" (Harvard Business School Ph.D. thesis, Baker Library, 1987).

Forbes, M., "Economic Value to the Customer" *McKinsey & Company Staff Paper* (February, 1979).

Gardner, D. M., "The Role of Price in Consumer Choice," in *Selected Aspects of Consumer Behavior: A Summary of the Perspective of Different Disciplines* (Washington, DC, National Science Foundation, 1978), pp. 415–434.

Gardner, D. M., Winer, F. W., *Proceedings of the 11th Paul D. Converse Symposium* (Chicago, American Marketing Association, 1982).

Geruson, R. J., "Elements of Strategy in IBM's Price and Product Behaviour" (Oxford University M.Phil. thesis, Bodliean Library, 1982).

Ghemewat, P., "Sustainable Advantage," *Harvard Business Review*, 64, No. 5 (September–October, 1986), pp. 53–58.

Gould, J. P., Sen, S. K., eds., *Pricing Strategy: Proceedings of a Conference, The Journal of Business*, 57, No. 1, Pt. 2 (Chicago, University of Chicago Press, January, 1984).

Grether, E. T., "Chamberlin's Theory of Monopolistic Competition and the Literature of Marketing," in E. Kuenne, ed., *Monopolistic Competition Theory: Studies in Impact. Essays in Honour of Edward H. Chamberlin* (New York, Wiley, 1967), pp. 307–328.

Howard, W. C., *Selling Industrial Products* (New York, Norton, 1973a).

Howard, W. C., "How Industry Buys," available from Harvard Business School Case Services, order number 3-574-083 (1973b).

Hedley, B., "A Fundamental Approach to Strategy Development," *Long Range Planning* (December, 1976), pp. 2–11.

Hedley, B., "Strategy and the 'Business Portfolio'," *Long Range Planning*, 10 (February, 1977), pp. 9–15.

Kuenne, R. E., *Monopolistic Competition Theory: Studies in Impact. Essays in Honor of Edward H. Chamberlain* (New York, John Wiley & Sons, 1967).

Lambert, D. R., "On Paying Homage to a False God: Comments on the Theory of the Firm's Role" in

O. C. Ferrell, S. Brown, C. Lamb, eds., *Conceptual and Theoretical Developments in Marketing* (Chicago, American Marketing Association, 1979), pp. 363–373.

Levitt, T., *Marketing for Business Growth,* 2d ed. (New York, McGraw Hill, 1974).

Levitt, T., *The Marketing Mode: Pathways to Corporate Growth* (New York, McGraw-Hill, 1982).

Levitt, T., *The Marketing Imagination* (New York, Free Press, 1983).

Machlup, F., "Characteristics and Types of Price Discrimination," in *Business Concentration and Price Policy* (Princeton, Princeton University Press, 1955), pp. 400–423.

Magnet, M., "How to Compete with IBM" (*Fortune,* 1984) in B. P. Shapiro, R. J. Dolan, J. A. Quelch, *Marketing Management Readings: From Theory to Practice,* III (Homewood, IL, Irwin, 1985), pp. 252–260.

Mantell, L. H., Sing, F. P., *Economics for Business Decisions* (New York, McGraw Hill, 1972).

Michaels, R., "Hedonic Prices and the Structure of the Digital Computer Industry," *The Journal of Industrial Economics,* 27, No. 3 (1979), pp. 263–275.

Mintzberg, H., "Patterns in Strategy Formation," *Management Science,* 24, No. 9 (May, 1978), pp. 934–948.

Mintzberg, H., "An Emerging Strategy of Direct Research," *Administrative Science Quarterly,* 24 (December, 1979), pp. 582–589.

Morierty, R. T., *Industrial Buying Behavior* (Lexington, MA, Lexington Books, 1983).

Nagle, T., "Economic Foundations for Pricing," *The Journal of Business,* 57, No. 1, Pt. 2 (January, 1984), pp. 3–26.

Nelson, P., "Information and Consumer Behavior" *Journal of Political Economy,* 78 (March–April, 1970), pp. 311–329.

Nelson, P., "Advertising as Information Once More," in D. Tverck, ed. *Issues in Advertising: The Economics of Persuasion* (Washington, DC, American Enterprise Institute, 1978), pp. 133–160.

Peterson, R. D., "Product Differentiation, Implicit Theorizing and the Methodology of Industrial Organization," *Nebraska Journal of Economics and Business,* 19 (Spring, 1980), pp. 22–36.

Phillips, L. W., Chang, D. R., and Buzzell, R. D., "Product Quality, Cost Position and Business Performance: A Test of Some Key Hypothesis," *Journal of Marketing,* 47 (Spring, 1983), pp. 26–43.

Porter, M. E., "How Competitive Forces Shape Strategy," *Harvard Business Review* (March–April, 1979), pp. 137–145.

Porter, M. E., *Competitive Strategy: Techniques for Analyzing Industries and Competitors* (New York: Free Press, 1980).

Porter, M. E., "The Contribution of Industrial Organization to Strategic Management: A Promise to be Realized," *Academy of Management Review* (October, 1981).

Porter, M. E., "Industrial Organization and the Evolution of Concepts for Strategic Planning: The New Learning," in T. H. Naylor, *The Integration of Corporate Planning Models and Economics* (Amsterdam, North Holland Publishing, 1982).

Porter, M. E., *Competitive Advantage: Creating and Sustaining Superior Performance* (New York, Free Press, 1985).

Quinn, J. B., "Strategic Change: Logical Incrementalism," *Sloan Management Review* (Fall, 1978), pp. 7–20.

Ratchford, B. T., Ford, G. T., "A Study of Prices and Market Shares in the Computer Mainframe Industry," *Journal of Business,* 49 (April, 1976), pp. 194–218.

Ratchford, B. T., Ford, G. T., "A Study of Prices and Market Shares in the Computer Mainframe Industry: A Reply," *Journal of Business,* 52, No. 1 (1979), pp. 125–134.

Robert, M. M., "Managing Your Competitor's Strategy," *Journal of Business Strategy* (March–April, 1990).

Samuelson, P. A., "The Monopolistic Competition Revolution, in R. Kuenne, ed., *Monopolistic Competition Theory: Studies in Impact. Essays in Honor of Edward H. Chamberlin* (New York, Wiley, 1967), pp. 103–138.

Scherer, F. M., *Industrial Market Structure and Economic Performance* (Chicago, Rand McNally, 1980).

Selected Aspects of Consumer Behavior: A Summary From the Perspective of Different Disciplines (Washington, DC, National Science Foundation, 1978).

Shapiro, B. P., "Can Marketing and Manufacturing Coexist?" *Harvard Business Review*, 56 (September–October, 1977), pp. 104–114.

Shapiro, B. P., *Industrial Product Policy: Managing the Existing Product Line* (Cambridge, MA, Marketing Science Institute, 1979).

Shapiro, C., "Premiums for High Quality Products as Returns to Reputations," *The Quarterly Journal of Economics* (November, 1983), pp. 659–679.

Sharpe, W. F., *The Economics of Computers* (New York, Columbia University Press, 1969).

Shepard, W. G., *Market Power and Economic Welfare* (New York, Random House, 1970).

Skeoch, L. A., *Canadian Competition Policy: Proceedings of a Conference Held at Queens University, Kingston, Ontario*, 20–21 January, 1972 (Kingston, Industrial Relations Centre of Queens University, 1972).

Smith, W. R., "Product Differentiation and Market Segmentation as Alternative Marketing Strategies," *Journal of Marketing* (July, 1956), pp. 3–8.

Smith, W. R., "Product Differentiation and Market Segmentation: Another Look," in D. M. Gardner, F. W. Winer, eds., *Proceedings of the 11th Paul D. Converse Symposium* (Chicago, American Marketing Association, 1982), pp. 1–9.

Sobel, R., *IBM: Colossus in Transition* (New York, Truman Talley, 1981).

Soma, J. T., *The Computer Industry: An Economic-Legal Analysis of its Technology and Growth* (Lexington, MA, Lexington Books, 1976).

Spekman, R. E., Wilson, D. T., *Issues in Industrial Marketing: A View to The Future* (Chicago, American Marketing Association, 1982).

Stern, L. W., Grabner, J. R., Jr., *Competition in the Marketplace* (Glenview, IL, Scott, Foresman and Company, 1970).

Stigler, G. J., "The Economics of Information," *Journal of Political Economy*, No. 3 (June, 1969), pp. 213–225.

Taylor, J. W., "The Role of Risk in Consumer Behaviour: A Comprehensive and Operational Theory of Risk Taking in Consumer Behaviour," *Journal of Marketing*, 38 (April, 1974), pp. 54–60.

Telex v. IBM, Northern District of Oklahoma. (Available through the Computer Industry Association Document Services, Encino, CA.)

United States v. IBM, Civil Action No. 69, Civ. 200 (D.N.E.), United States District Court of the Southern District of New York, filed 17 January 1969, dismissed 8 January 1981.

Weil, U., *Information Systems in the 80s: Products, Markets, Vendors* (Englewood Cliffs, NJ, Prentice Hall, 1982).

White, P. A., *The Dominant Firm: A Study in Market Power* (Ann Arbor, Michigan, UMI Research Press, 1983).

Wilde, L., "The Economics of Consumer Information Acquisition," *Journal of Business*, 53, No. 3 (July, 1980), pp. 143–158.

Winter, F. W., "Market Segmentation: A Review of its Problems and Promise," in D. M. Gardner, F. W. Winter, eds., *Proceedings of the 11th Paul D. Converse Symposium* (Chicago, American Marketing Association, 1982), pp. 19–29.

Index